Four-coupled Tank Locomotive Classes Built by the Great Western Railway

Front cover photo:
The preserved 4866 in GW livery at the Great Western Railway Centre, Didcot, 2017. (David Maidment)

Back cover photos:
Broad gauge Hawthorn class 2-4-0ST *Ostrich* rebuilt from 1865 built 2-4-0 in 1877, c1890. (LPC/F.K. Davies/John Hodge Collections)

1105 in BR livery of black with red backed numberplate at Danygraig shed, 1953. (N. Harrop/MLS Collection)

The Wolverhampton built prototype 0-4-2T of the 517 class, in the blue/green livery applied in the 19th century to the Wolverhampton constructed engines. It is seen at Birmingham Snow Hill in July 1900, the original photograph colourised by Andrew Knowles of the Great Western Society to illustrate this rarely seen livery. (G.M.Perkins/Steve Armitage Collection/Rail-Online)

Four-coupled Tank Locomotive Classes Built by the Great Western Railway

DAVID MAIDMENT

First published in Great Britain in 2023 by
Pen & Sword Transport
An imprint of Pen & Sword Books Ltd
Yorkshire - Philadelphia

Copyright © David Maidment, 2023

ISBN 978 1 39902 256 9

The right of David Maidment to be identified as author of this work has been asserted by him in accordance with the Copyright, Designs and Patents Act 1988.

A CIP catalogue record for this book is available from the British Library.

All rights reserved. No part of this book may be reproduced or transmitted in any form or by any means, electronic or mechanical including photocopying, recording or by any information storage and retrieval system, without permission from the Publisher in writing.

Typeset in Palatino by SJmagic DESIGN SERVICES, India.
Printed and bound by Printworks Global Ltd, London/Hong Kong.

Pen & Sword Books Ltd incorporates the Imprints of Pen & Sword Books Archaeology, Atlas, Aviation, Battleground, Discovery, Family History, History, Maritime, Military, Naval, Politics, Railways, Select, Transport, True Crime, Fiction, Frontline Books, Leo Cooper, Praetorian Press, Seaforth Publishing, Wharncliffe and White Owl.

For a complete list of Pen & Sword titles please contact:

PEN & SWORD BOOKS LIMITED
47 Church Street, Barnsley, South Yorkshire, S70 2AS, England
E-mail: enquiries@pen-and-sword.co.uk
Website: www.pen-and-sword.co.uk

Or

PEN AND SWORD BOOKS
1950 Lawrence Rd, Havertown, PA 19083, USA
E-mail: Uspen-and-sword@casematepublishers.com
Website: www.penandswordbooks.com

All royalties from this book will be donated to the Railway Children charity [reg. no. 1058991] [www.railwaychildren.org.uk]
Other books by David Maidment:

Novels (Religious historical fiction)
The Child Madonna, Melrose Books, 2009
The Missing Madonna, PublishNation, 2012
The Madonna and her Sons, PublishNation, 2015
The Reluctant Traitor, PublishNation, 2021

Novels (Railway fiction)
Lives on the Line, Max Books, 2013
Steamy Stories, PublishNation, 2021 (Short stories)

Non-fiction (Railways)
The Toss of a Coin, PublishNation, 2014
A Privileged Journey, Pen & Sword, 2015
An Indian Summer of Steam, Pen & Sword, 2015
Great Western Eight-Coupled Heavy Freight Locomotives, Pen & Sword, 2015
Great Western Moguls and Prairies, Pen & Sword, 2016
Southern Urie and Maunsell 2-cylinder 4-6-0s, Pen & Sword, 2016
Great Western Small-Wheeled Double-Framed 4-4-0s, Pen & Sword, 2017
The Development of the German Pacific Locomotive, Pen & Sword, 2017
Great Western Large-Wheeled Double-Framed 4-4-0s, Pen & Sword, 2017
Great Western Counties, 4-4-0s, 4-4-2Ts & 4-6-0s, Pen & Sword, 2018
Southern Maunsell Moguls and Tank Engines, Pen & Sword, 2018
Southern Maunsell 4-4-0s, Pen & Sword, 2019
Great Western Granges, Pen & Sword, 2019
Cambrian Railways Gallery, Pen & Sword, 2019
Great Western Panniers, Pen & Sword, 2019
Great Western Kings, Pen & Sword, 2020
Great Western & Absorbed Railway 0-6-2Ts, Pen & Sword, 2020
Drummond's L&SWR Passenger & Mixed Traffic Locomotives, Pen & Sword, 2020
Southern 0-6-0 Tender Locomotives, Pen & Sword, 2021
LNER 4-6-0 Locomotives, Pen & Sword, 2021
Midland & LMS 4-4-0s, Pen & Sword, 2021
Great Western Castle 4-6-0 Locomotives, 1923-1959, Pen & Sword, 2022
Great Western Castle 4-6-0 Locomotives, The Final Years, 1960-1965, Pen & Sword, 2022
Great Western Castle 4-6-0 Locomotives in the Preservation Era, Pen & Sword, 2023

Non-fiction (Street Children)
The Other Railway Children, PublishNation, 2012
Nobody ever listened to me, PublishNation, 2012

CONTENTS

	Acknowledgements	6
	Introduction	7
Chapter 1	**The Engineers**	9
	Sir Daniel Gooch	9
	Joseph Armstrong	9
	George Armstrong	10
	William Dean	10
	G.J. Churchward	11
	C.B. Collett	11
Chapter 2	**The Broad Gauge Locomotives**	13
Chapter 3	**Standard Gauge Locomotives built by the GWR before 1923**	22
	Wolverhampton designed classes	25
	Swindon designed classes	46
	Experimental locomotives	87
Chapter 4	**Locomotives built by the GWR after 1923**	94
Chapter 5	**Preservation**	136
	Colour Section	145
	Appendix	161
	Bibliography	195
	Index	196

ACKNOWLEDGEMENTS

I could not have written this book without access to the details of the large variety of four-coupled tank engines that belonged at one time or another to the Great Western Railway or its constituent parts without the use of the comprehensive collection of books and photographs held by the Manchester Locomotive Society at their clubrooms on Stockport railway station. I am particularly grateful to the copies of the RCTS research on Great Western locomotives held in the MLS library and the archive of photographs made available to me, free of any publication fee, by their photo archivist, Paul Shackcloth, as once again I am donating all royalties from this book to the Railway Children charity (www.railwaychildren.org.uk) which I founded in 1995. The charity supports street children living on the railway and bus stations of India and East Africa and works with the British Transport Police to protect and counsel runaway children picked up on the railway stations of our own country. I also acknowledge the help given me by a colleague for many years in the railway industry, John Hodge, who has made a large number of photographs available to me from the F.K. Davies collection which he owns. As is often the case with collections, it is not always possible to identify the original photographer and if I have missed giving credit to the copyright holder, please contact the publisher. I am also grateful to Laurence Waters of the Great Western Railway Trust who has provided colour photographs from the Trust collection to augment the few I have been able to take myself on visits to the GW Centre at Didcot.

I also thank the Pen & Sword company staff for their usual helpful and very competent way in which they have brought this book to publication – my editor, Carol Trow, Commissioning Editor and friend, John Scott-Morgan, Transport Production Manager Janet Brookes and the design and marketing staff at the company headquarters at Barnsley – or during this difficult period of lockdown from the Covid pandemic, working from home.

David Maidment
March 2023

INTRODUCTION

I didn't realise when I agreed to tackle the story of the Great Western Railway's four-coupled tank engines just how many there were. My immediate thought and inspiration was to tell the history of the GW's classic branch line engines, the 14XX 0-4-2 tank engines and their predecessors, the Wolverhampton built '517' class 0-4-2Ts and the Swindon built 2-4-0Ts – the 'Met Tanks'. Then, as I began to research in the library and photo archives of the Manchester Locomotive Society in their clubrooms on Stockport station, the scale of the work that I had committed myself to in signing the contract with Pen & Sword Books began to dawn on me.

Delving back into the nineteenth century, the number of small railway companies that existed in the West of England, the West Midlands and South Wales that struggled and were taken over by the Great Western surprised me and their stories were complex. All had four-coupled locomotives within their fleets almost from the beginning and many had comparatively short routes so tank engines were appropriate for nearly all of them. Most of these companies were taken over by the Great Western towards the end of the nineteenth century, a few retained their nominal existence, but their operations were carried out by the GWR, and others, most notably the majority in South Wales and the Midland & South Western Junction Railway, were not absorbed by the GWR until 1922. The story of these 'absorbed' engines is recounted in the companion volume *Four-coupled tank locomotives absorbed by the Great Western Railway*.

In lines in South Wales and South Devon and Cornwall and in the London suburban area, four-coupled tank engines became for decades the main passenger engines and only as traffic levels grew towards the turn of the century did the need for larger locomotives relegate the four-coupled varieties to secondary and branch line work. As well as the passenger work, local freight trip work and shunting operations also became the domain of four-coupled tank engines, many remaining to near the end of steam in industrial sidings, collieries and docks. And, on the country branches, many engines, some dating from the 1850s and 1860s, though often much modified or rebuilt, lived on to a great age – at least until the 1920s and '30s, when Charles Collett, in his drive for cost reduction, replaced many by the simple, but very similar 48XX (later 14XX) 0-4-2 tank engines which remained the staple power of Great Western branch lines along with the 0-6-0 pannier tanks until the lines were closed or steam power was replaced by diesel multiple units. The Great Western was also among the railways that pioneered the use of railmotor vehicles incorporating a steam engine and carriage on the same chassis and as the locomotive part of the railmotor was of the 0-4-0T wheel arrangement, I have included the designs of both the GW and constituent companies.

I therefore tackle in the two books a comprehensive review of all the Great Western's four-coupled tank engines from the main company itself and all its constituent companies that merged or were taken over by it. I start with a chapter about the Broad Gauge engines, with following chapters on engines constructed by the Great Western Railway itself, both before and after the 1923 Grouping. I include, as is my practice, my own somewhat limited experience with these engines – regrettably just the 14XX and one journey in the late 1950s to Swansea Docks when I surprised the foremen at Swansea East Dock and Danygraig by turning up with a shed permit before 6am and the shunting

0-4-0STs set off for their morning shift on the docks. I conclude with a description of preserved locomotives.

There is inevitably some duplication with earlier books that I have written for Pen & Sword. I have given a brief description of many of the engineers in previous books, but I repeat them for the sake of those readers who have only purchased this book. The '2221 County Tanks' and the '4600' 4-4-2T were included in my book on the GW Counties and a few of the engines described in this book were rebuilt with pannier tanks and were included in my Pen & Sword's *Great Western Pannier Tank Classes*. I repeat the text describing the design, construction and operation of these classes, but have only included a few photos for completeness, the 'County Tanks' in particular being much more widely illustrated in the previous book.

Chapter 1
THE ENGINEERS

I have included brief biographies of William Dean, George Jackson Churchward and Charles Collett of the Great Western, who were responsible for the design of a number of the four-coupled tank locomotives – it is a repeat of what I have published earlier in my GW Pen & Sword books, but I have included it again for the benefit of new readers to the series.

Sir Daniel Gooch
Daniel Gooch was born in 1816 in Bebington, Northumberland, the son of an ironfounder and his family who moved to Tredegar in 1831. He trained under Thomas Ellis who worked with Samuel Homfray and Richard Trevithick to pioneer steam locomotion. At the age of twenty he was recruited by Brunel as Superintendent of Locomotive Engines, starting in 1837. In 1840 he found the site for Swindon Works and in 1846 designed the prototype of the 'Iron Duke' broad gauge 4-2-2, *Great Western*, the first engine constructed at the new Works. Although he was mainly involved in the design and construction of broad gauge engines at Swindon, between 1854 and 1864 he designed a number of standard gauge engines for the GWR's Northern Division at Wolverhampton, and this included the first 0-6-0 saddle tanks, the origin of the long line of GW saddle and pannier tanks (Nos.93 and 94).

He resigned in 1864 when he entered politics as a Conservative MP but continued as a member of the GWR Board, a post he retained until 1889. He died on 15 October 1889.

Joseph Armstrong
Joseph Armstrong was born in Bewcastle, Cumberland, in 1816 and lived with his family in Newburn-on-Tyne from 1824. He attended Bruce's School in Newcastle where Robert Stephenson had studied and would have been well aware of the Wylam Waggonway at Newburn where the famous locomotive *Puffing Billy* was operating. His first employment was at the local Walbottle Colliery railway worked by stationary engines. He gained some experience driving locomotives on the Stockton and Darlington Railway through the influence of Stephenson and Timothy Hackworth, and at the age of twenty in 1836 became a driver on the Liverpool and Manchester Railway. He was appointed as a foreman on the Hull and Selby Railway and in 1847 he was promoted to assistant locomotive superintendent on the Shrewsbury and Chester Railway, becoming its locomotive superintendent in 1853, taking responsibility for locomotives of the Shrewsbury and Birmingham Railway at the same time, working at the Wolverhampton Locomotive Works.

In September 1854, the two railways amalgamated with the GWR as its Northern Division, Joseph Armstrong remaining the Northern Division Locomotive Engineer reporting to Daniel Gooch at Swindon. In 1859, Wolverhampton began constructing locomotives to Armstrong's design under the delegated authority of Gooch. In 1864, Gooch resigned and Joseph Armstrong replaced him, now as Locomotive, Carriage & Wagon Superintendent. He moved to Swindon leaving Wolverhampton in charge of his previous assistant, his younger brother, George. He was hardworking and strict, intolerant of corruption and injustice and involved in civic affairs. He was a Methodist lay preacher and generous to his staff and townspeople, president of the Swindon Mechanics Institute and Chairman of the Swindon New Town Board.

After the Gauge Commission decided in 1846 against the expansion of the broad gauge, Armstrong was faced with designing engines to replace them and from 1868 onwards was building engines entirely for standard and mixed gauge track. He had to cover all motive power needs from express power to shunting locomotives and many of the saddle tanks to be described in this book originated in his era. He was specifically responsible for the design of the '302' class and his 'masterpiece' tank engine, the '1076' or 'Buffalo' class. Many other classes were built at Wolverhampton during this period under the aegis of his younger brother, George, and it is not possible to distinguish the extent of the influence Joseph had over his brother's designs. It is said that he left the Great Western better provided with sound engines for every class of traffic than any other railway in Britain, and probably in the world.

He died of a heart attack in 1877 aged 61 and his funeral in June was attended by 2,000 workers from Swindon Works, 100 from Wolverhampton and many from other railway towns on the GWR and other railways, such was his reputation – in all 6,000 people crowded into St Mark's churchyard at Swindon where he was buried. He was survived by his wife, Sarah, whom he had married in 1848, and nine children, four of whom were apprenticed at Swindon Works.

George Armstrong

George Armstrong was born in Canada in 1822 but moved with his family to Bewcastle in 1824. He too moved to Newcastle and followed his older brother in his interests in engineering around the Northumberland colliery early railways. He also worked at Walbottle Colliery and followed his brother to the Hull and Selby Railway and worked for a period on the Northern Railway in France. George returned to Britain in 1848 and became an engine driver, then foreman on the Shrewsbury and Chester Railway and followed his brother again to Wolverhampton, becoming Joseph's assistant and Works Manager.

When Joseph Armstrong moved to Swindon in 1864, William Dean became Joseph's assistant at Swindon and George was promoted as the GWR Northern Division Locomotive Superintendent at Wolverhampton, where his brother gave him a free hand in designing standard gauge locomotives. George built relatively few tender engines, but significant numbers of 2-4-0 and 0-6-0 side and saddle tanks. His four-coupled saddle tanks are the subject of this book, many of the 0-6-0 variety being converted during the Churchward and Collett eras to the pannier tank form (the class '1016' double-framed saddle tanks, the '645', '850' classes and the rebuilding of 0-6-0 tender engines as saddle tanks, of classes '119' and '322').

George remained in charge of the Wolverhampton Works for thirty-three years, during which time he managed the building or rebuilding of 1,139 locomotives. He retired in 1897 and died in 1901. He remained single and is buried at St Mary's church, Bushbury, Wolverhampton.

William Dean

William Dean was born in 1840, educated at Haberdasher's School in New Cross and in 1855 he was apprenticed to Joseph Armstrong, then Locomotive Superintendent of the Great Western at Stafford Road Works, Wolverhampton. He advanced rapidly, becoming Armstrong's chief assistant when still only 23 years of age. Armstrong moved to Swindon as Chief Locomotive, Carriage and Wagon Superintendent in 1864 and left Dean in charge of Wolverhampton Works. Dean moved to Swindon in 1868 and became Chief Assistant Superintendent there until Armstrong's early death in 1877.

Dean was appointed in his place at the age of 37 and held office for exactly twenty-five years and had over 13,000 men under him at Swindon itself plus enginemen all over the GWR system. He became a Justice of the Peace there and was highly respected both by his staff and in the community as a caring and public-spirited manager. Already since 1868 a full member of the Institute of Mechanical Engineers, in 1878 he attained a similar rank in the Institute of Civil Engineers. He was a very practical man and aimed for simplicity, economy and easy maintenance of his rolling stock. He added considerably to the growing stock of the company's 0-6-0 saddle tanks, designing and constructing the '1813', '1661', '1854' and '655' classes and the single 4-4-0 pannier tank, 1490.

By the 1890s his health had begun to deteriorate and by 1896 his mental health began to crumble. Churchward was appointed as his assistant in 1897 and had the very delicate task of supporting him during his final years in theoretical charge, as the Company, after so many excellent years of service, was reluctant to terminate Dean's career. He eventually retired in

1902 aged 63, much revered in Swindon despite his failing health, and moved to a house in Folkestone bought for him by the GWR Company. He died in 1905 aged 66.

George Jackson Churchward
Churchward, Dean's successor and virtual co-manager during the final five years or so, was born in 1857 in Stoke Gabriel on the River Dart between Kingswear and Totnes, and joined the South Devon Railway at Newton Abbot in 1873. After absorption of that railway by the Great Western in 1876, he transferred aged just nineteen to the Swindon Drawing Office and, after a few rapid promotions, was appointed as Carriage and Wagon Works Manager in 1885. Ten years later he became Swindon Works Manager and identified as Dean's successor when he became his Chief Assistant in 1897. Although he was not appointed as Locomotive Superintendent until 1 June 1902, he had been developing his ideas within the ample scope given him by Dean, and had already written a paper on a scheme for a limited number of 'standard' locomotive designs by January 1901. However, in the interim he maintained a steady production of Dean-designed engines, albeit showing an increasing influence of his own ideas, especially boiler design. The '2721' and smaller wheeled '2021' class emerged during the dual management period, developed from Dean's earlier designs.

Although he is remembered for his main line standard express passenger and freight engines, his development of boilers and use of the Belpaire firebox meant that he initiated on a large scale the reboilering and equipping with pannier tanks many of his predecessors' saddle tanks, although he built no new shunting engines of his own. He did adapt one of his predecessor's designs to create an 0-6-4 crane tank for service activity in both Swindon and Wolverhampton works and drew the outline plan for an 0-8-0 pannier tank, but this was never developed.

Churchward had an even temperament and a dignified bearing suggesting a 'country squire', strengthened by his interest in country pursuits. But he was also a good administrator and leader of men. He drew out the best from his staff and created a culture of good teamwork, a tradition and practice he inherited from Dean and his predecessors. In 1916 his title was changed to that of Chief Mechanical Engineer, he was awarded the CBE at the end of the war and in October 1920 he was the first Honorary Freeman of Swindon, of which he'd been the first Mayor as far back as 1900. It is well known that his life ended run down by one of his successor's engines whilst crossing the line from his home to the Works, nearly twelve years after his retirement.

Charles B. Collett, GWR, 1922-41
Charles Benjamin Collett was born on 10 September 1871, the son of a journalist. After an initial post in a marine engineering firm, he went to Swindon in December 1893 as a draughtsman and was spotted by Churchward and appointed as Assistant Manager in the Locomotive Works in 1900, and Works Manager in 1912, becoming Churchward's deputy in 1919. The far-seeing strategy of Churchward was the foundation of Collett's design programme and his involvement in the management of the Works during the production of the standard GW locomotives gave him a clear insight into the background of the basic designs, which he continued from his appointment to the senior post in 1922.

The Board required him to reduce annual costs by £500,000, a colossal challenge, in excess of £25 million in today's currency. Some workers were laid off, but Collett sought savings from improved methods in the Works and keeping most essential new builds to existing designs. The pressure on Collett was to reduce costs. He continued to refine Churchward's standardisation policy, reducing the seventeen locomotive types in fifty-two classes in 1921 to thirteen basic types in thirty-seven classes by his retirement. His cost-cutting standardisation measures included the design of the 56XX to replace the myriad Victorian 0-6-0 and 0-6-2 saddle tanks of the absorbed South Wales railways and the 48XX 0-4-2Ts to replace the increasingly costly four-coupled branch line engines, some dating back to the middle of the nineteenth century.

Collett was a very different character to Churchward. Although he had had a privileged initial upbringing in the family home of Grafton Manor, he had lost his elder brother when he was only ten, then his father died in 1884 when he was just thirteen and a pupil at Merchant Taylors' public school. He was, perhaps because of this, a more private person, devoted to Ethelwyn, his wife, the daughter of a Bloomsbury clergyman – Churchward was a bachelor – and he was greatly affected by the tragedy of her premature death after a short illness at the age of forty-seven in 1923. He became harder to approach and understand. He was a fair man and competent, though introverted and increasingly withdrawn, obsessed in his later years with spiritualism in an urge to retain some sort of contact with his beloved wife. He was never greatly involved in civic affairs though he was a magistrate for a number of years.

Collett was reluctant to retire and remained in position until he was seventy, believing that he had been encouraged to stay by Viscount Churchill, a former Chairman of the Board. Although he hated war and violence, even feeling little sympathy for Churchward's enthusiasm for blood sports, he was at the helm as the Second World War grew nearer and was pressed to make much of Swindon Works available to munitions production and resisted, remembering the toll on the GW locomotive fleet that the 1914-18 war effort had caused, for which he'd received the OBE. His reluctance in 1939 to commit Swindon Works to the war effort caused him even to be investigated initially by MI5, though the Board over-ruled his decision to keep the Works for locomotive construction and maintenance. He was 'persuaded' to make way for his assistant, Hawksworth. He retired to Wimbledon and died in 1952, aged eighty.

Chapter 2
THE BROAD GAUGE LOCOMOTIVES

The Great Western main lines constructed from 1835 onwards on Brunel's 7ft broad gauge were copied by a number of smaller companies in the same geographical area. The first GW broad gauge four-coupled tank locomotives were 2-4-0 tank locomotives converted from 2-4-0 tender engines and appeared as early as 1841 and similar designs to the GW engines were purchased or built by engineering companies for the South Devon, Bristol & Exeter and Torbay & Brixham railways in the West of England and the Vale of Neath, Llynvi & Ogmore and South Wales Mineral railways in Wales. Information about some of the earliest locomotives is sketchy and I have been very reliant on information unearthed by the Railway Correspondence & Travel Society and documented in their comprehensive reviews of Great Western locomotives published in the early 1950s. At the end of Gooch's reign as Locomotive Superintendent in 1864, fifty-five of the Great Western's 371 locomotives were 2-4-0 tank locomotives. By Armstrong's appointment to follow Gooch, the writing was already on the wall for the broad gauge and only seventy more engines for this gauge exclusively were to be built of which ten were 2-4-0 saddle tanks converted from his 'Hawthorn' class of 2-4-0 tender engines. As conversion to standard gauge rolled out, William Dean in the 1880s built some 'convertibles', engines designed to run initially on the broad gauge, but were capable of conversion relatively easily to run on the 4ft 8½in railway.

GWR classes

'Leo' class 2-4-0T, 1841

Eighteen 2-4-0 tender engines were designed by Gooch and built by three different manufacturers, R. & W. Hawthorn & Co. of

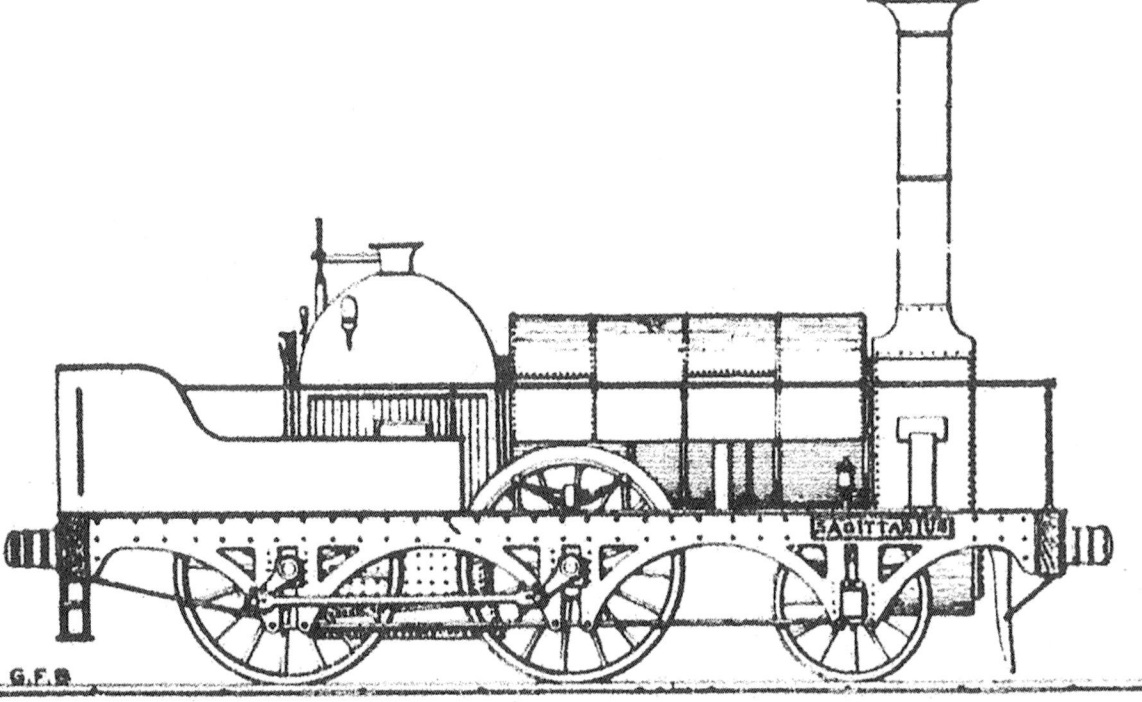

Drawing of 'Leo' class 2-4-0ST broad gauge locomotive *Sagittarius*. (G.F. Bird/LCGB)

Newcastle, Fenton, Murray & Jackson of Leeds and Rothwell & Co. of Bolton, in 1841 and 1842. They were intended as goods engines. They were given names, not numbers, and were known as the 'Leo' class. All eighteen were subsequently rebuilt as 2-4-0 saddle tanks to increase the adhesion needed for freight operation. Their key dimensions were:

- Coupled wheels: 5ft 0in
- Leading wheels: 3ft 6in
- Cylinders: 15in x 18in
- Boiler pressure: 50lb psi
- Grate area: 11.5sqft
- Heating surface: 467sqft
- Weight: 25 tons 14 cwt
- Axleload: 10½ tons

Their names were:

Elephant	Buffalo	Dromedary
Hecla	Stromboli	Etna
Aries	Taurus	Gemini
Cancer	Leo	Virgo
Libra	Scorpio	Sagittarius
Capricornus	Aquarius	Pisces

They were used widely over the system, including operations over the South Devon Railway, and in the 1860s some were used on the West London Railway. The first examples were taken out of traffic in the mid-1860s, and most were withdrawn between 1870 and 1872, the last survivors being *Cancer* and *Pisces* withdrawn in June 1874. The average class mileage was around 400,000.

'Corsair' class 4-4-0ST, 1849

Two 4-4-0 tank engines designed by Daniel Gooch, also known as the 'Bogie' class and built at Swindon Works in 1849, entered the service named *Corsair* and *Brigand*. Their key dimensions were:

- Coupled wheels: 6ft 0in
- Bogie wheels: 3ft 6in
- Cylinders: 17in x 24in
- Boiler pressure: 60lb psi
- Heating surface: 1,255.7sqft
- Grate area: 19sqft
- Weight: 35 tons 15 cwt
- Axleload: 10 tons 12½ cwt
- Tank capacity: 930 gallons

A further thirteen engines of similar design were built by R. & W. Hawthorn in 1854 and 1855. Their coupled wheel dimension was reduced to 5ft 9in and the grate area was slightly smaller at 18.44sqft. They were named:

Sappho	Homer	Virgil
Horace	Ovid	Juvenal
Seneca	Lucretius	Theocritus
Statius	Euripides	Hesiod
Lucan.		

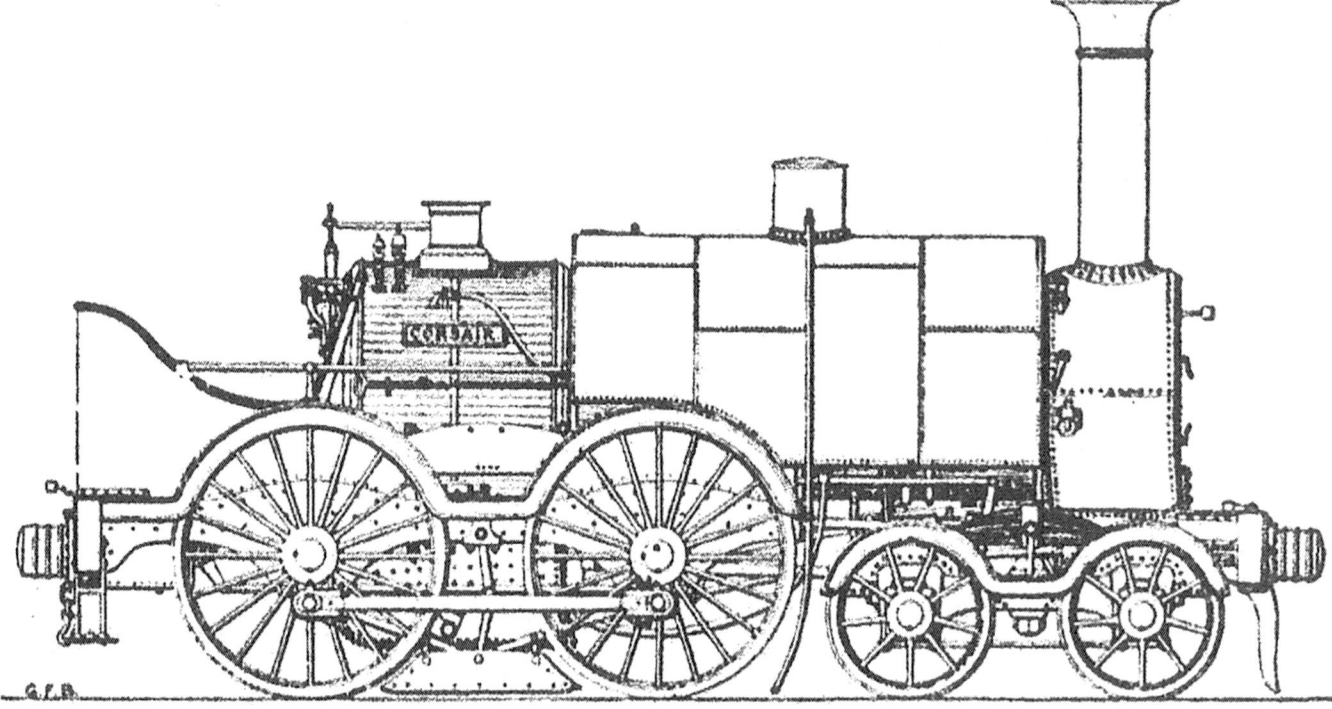

Drawing of 'Bogie' class 4-4-0ST broad gauge locomotive, named *Corsair*. (G.F. Bird/LCGB)

It would appear that a classical education was necessary for employment in the GWR's motive power department! They were saddle tanks with inside sandwich frames extending from the front of the coupled wheels to the back bufferbeam. The boiler barrel connected the cylinders and the main frame. The bogie swivelled on a ball and socket joint riveted to the boiler barrel and steam chest. The 1848 drawings depicted a tender engine design, and it was the intention to use them on the South Devon Railway for which at that time the GWR was contracted to provide the power. The two Swindon built engines were based at Plymouth until 1852. Both the Bristol and Exeter and South Devon railways provided their own locomotive power by 1852 and the GW 4-4-0Ts were then based at Bristol, South Wales or Swindon.

The engines were withdrawn between 1871 and 1873 with the exception of the Hawthorn built *Horace* which survived until 1880, ending its days in Devon and Cornwall. The class averaged around 300,000 miles in their 16-19 year life, the two Swindon built engines lasting nearly twenty-four years and *Horace*, over twenty-six. Six were sold to iron works or collieries after their withdrawal.

'Metropolitan' class 2-4-0T, 1862

Twenty-two 2-4-0 outside cylinder tank engines were built for working over the Great Western and Metropolitan lines between 1862 and 1864. Designed by Daniel Gooch, they were the first in the country to be fitted with condensing apparatus. Six were constructed by the Vulcan Foundry at Newton-le-Willows, another six by Kitson and Co. of Leeds and the final ten were built at Swindon. Their cylinders were inclined to clear the leading wheels. Their key known dimensions were:

Coupled wheels:	6ft 0in
Leading wheels:	3ft 6in
Cylinders:	16in x 24in
Grate area:	18sqft
Tank capacity	718 gallons

The Vulcan engines were named:

Hornet*	Bee	Gnat
Wasp	Mosquito	Locust

The Kitson engines were named:

Shah	Bey	Czar
Mogul*	Kaiser	Khan

The Swindon engines were named:

Fleur de Lis	Rose	Thistle
Shamrock	Camelia	Azalia*
Lily*	Myrtle*	Violet*
Laurel*		

Their initial work was in the London suburban area, but they were not particularly successful and most had their condensing gear removed and seven were converted as 2-4-0 tender engines (marked * above), the last four possibly before going into service. Four of the tender engines – *Hornet, Mogul, Myrtle* and *Laurel* – were transferred to Salisbury and the 2-4-0T *Czar* to Bristol. *Czar* was the first to be withdrawn in 1871 after a life of less than nine years and most were withdrawn between 1872 and 1877, the tank engines surviving the rebuilt tender engines, all of which had gone by 1873. The Kitson built engines seem to have been the least successful, for after *Czar*, four of the others were withdrawn in June 1872 and *Khan*, the last of the Kitson's, in December. The last survivor of the class was *Shamrock*, withdrawn

A very early photo of the broad gauge 2-4-0 'Metropolitan Tank' *Locust* built by the Vulcan Foundry in 1862. They were gone by the mid-1870s so this photo must be around 1870. (LPC/F.K. Davies/John Hodge Collections)

Above: **Broad gauge** Hawthorn class 2-4-0ST *Melling*, the first to be rebuilt from 1865 built 2-4-0 in 1877, withdrawn at the end of the broad gauge in May 1892. (LPC/F.K. Davies/John Hodge Collections)

Below: **Broad gauge** Hawthorn class 2-4-0ST *Ostrich* rebuilt from 1865 built 2-4-0 in 1877 and viewed from the other side, c1890. (LPC/F.K. Davies/John Hodge Collections)

in December 1877, after a life of just fourteen years. Their boilers were retained, however, for use on stationary work, one – that from *Lily* – for use on Brunel's *SS Great Eastern*.

'Hawthorn' class 2-4-0T, 1865

Joseph Armstrong designed the 'Hawthorn' class of 2-4-0 tender engines in 1865/6 and twenty were constructed by Slaughter, Gruning & Co. (name changed to Avonside Engine Co. later in 1865) and six at Swindon in 1865/6, which were classified as 'renewals'. Ten of the class were rebuilt as 2-4-0 saddle tanks in 1877 and these seven were built by Avonside:

Melling	*Roberts*	*Hedley*
Bury	*Beyer*	*Penn*
Stewart		

These three were converted from the six Swindon built 2-4-0s:

| *Ostrich* | *Cerberus* | *Pollux* |

After rebuilding, their key dimensions were:

Coupled wheels:	5ft 0in
Leading wheels:	3ft 6in
Cylinders:	17in x 24in
Boiler pressure	130lb psi
Heating surface:	1,201sqft
Grate area:	19sqft
Weight:	36 tons 4 cwt

Both the Bristol & Exeter Railway and the South Devon Railway were absorbed by the Great Western in 1876 and these ten saddle tanks were drafted to work west of Exeter. By the 1880s, they were used on the Devon branches operating from Newton Abbot and Plymouth

The Broad Gauge Locomotives • 17

Broad gauge Hawthorn class 2-4-0ST *Hedley* rebuilt from 1865 built 2-4-0 in 1877 which finished its life as a stationary boiler at Neath in 1929. It is seen here before withdrawal from active GW traffic, c1890. (LPC/F.K. Davies/ John Hodge Collections)

Dean's 'Convertible' 2-4-0 side tank No.3507, built in 1885 for the West of England broad gauge section west of Exeter, and converted in 1891 to a broad gauge 2-4-0 tender engine, c 1890. (LPC/F.K. Davies/ John Hodge Collections)

to Ashburton, Moretonhampstead, Kingswear and Launceston. All of the rebuilt 'Hawthorns' lasted to the end of the Broad Gauge in May 1892 apart from *Beyer* which was withdrawn in 1887. *Hedley* was retained to work a stone-crusher at Conwil Colliery in 1893 and was actually overhauled in 1905, becoming a stationary boiler at Neath Engineering Depot until condemned in 1914. It was not cut up until 1929 – at Swindon.

3501-3510 'Convertibles' class 2-4-0T, 1885

With the demise of the broad gauge in sight, William Dean constructed a number of locomotives for broad gauge use west of Exeter but with double frames providing outside bearings for the larger gauge and the ability to convert to the

18 • FOUR-COUPLED TANK LOCOMOTIVE CLASSES BUILT BY THE GREAT WESTERN RAILWAY

Dean's 'Convertible' 2-4-0 side tank No.3509, built in 1885 for the West of England broad gauge section until its conversion to standard gauge in May 1892 when it was rebuilt as a standard gauge tender engine. This was taken after withdrawal and before conversion, c1892. (GWR Official/F.K. Davies/John Hodge Collections)

standard gauge. These engines were known as 'Convertibles' and whilst most were 0-6-0 goods engines, some built by Joseph Armstrong as early as 1876, the first passenger 'Convertibles' were ten 2-4-0 side tanks, 3501-3510, built in 1885. Their key dimensions were:

Coupled wheels:	5ft 1in
Leading wheels:	3ft 6in
Cylinders:	17in x 26in
Boiler pressure:	140lb psi
Heating surface:	1,209.9sqft
Grate area:	15.2sqft
Weight:	43 tons 18 cwt
Axleload:	16½ tons
Tank capacity:	1,000 gallons

Five of these tank engines – 3501, 3502, 3505, 3507 and 3508 – were converted to tender engines in 1890 and 1891 to power the new *Cornishman* non-stop between Exeter and Plymouth and after the withdrawal of the other five tank engines on the abolition of broad gauge in May 1892, they too were converted to tender engines for standard gauge operation.

3541-3560 'Convertibles' class 0-4-2T, 1888, reb.0-4-4T, 1890

William Dean designed and Swindon Works constructed twenty 0-4-2 side tanks for Standard Gauge in 1887 and followed these with another twenty of similar basic design but as 'Convertibles' with saddle tanks in 1888 for use on the remaining broad gauge lines west of Exeter. They were numbered 3541-3560 but their initial running caused concern as they were unsteady and prone to derailment, so the final example, 3560, constructed in July 1889, was built to the 0-4-4T wheel arrangement and all the other nineteen saddle tanks were rebuilt as 0-4-4Ts similarly in 1890 and early 1891. After conversion of the lines west of Exeter in May 1892 the twenty locomotives were

The Broad Gauge Locomotives • 19

'Convertible' 3541 as a broad gauge 0-4-2 saddle tank built in 1888 before its conversion to an 0-4-4 side tank in April 1890. (LPC/F.K. Davies/John Hodge Collections

all converted to standard gauge, but their riding, even as 0-4-4Ts, continued to cause concern and they were rebuilt as 4-4-0 tender engines between 1899 and 1902 and gave satisfactory service for many years, most being withdrawn in the late 1920s with 3557 lasting until 1934. A full description of the rebuilds and their work is described in my earlier Pen & Sword book, *Great Western Small-wheeled Double-framed 4-4-0 Tender Locomotives*, published in 2017. The key dimensions of the 0-4-2STs were:

Coupled wheels:	5ft 0in
Trailing wheels:	4ft 0in
Cylinders:	17in x 24in
Boiler pressure:	180lb psi
Heating surface:	1,179.1sqft
Grate area:	17.2sqft
Weight:	46 tons 1 cwt

Axleload:	15 tons
Tank capacity:	1,130 gallons

Revised dimensions after rebuilding as 0-4-4 side tanks were:

Rear bogie wheels:	3ft 6in
Tank capacity:	1,075 gallons

Weight:	Not given but assumed to be increased.

The whole class operated west of Exeter on trains through to Penzance until the abolition of the broad gauge in 1892.

'Convertible' 3557 as a broad gauge 0-4-2 saddle tank built in 1888 before its conversion to an 0-4-4 side tank in February 1891. (LPC/F.K. Davies/John Hodge Collections)

'Convertible' 3560 built as a broad gauge 0-4-4 side tank in 1889 after the 0-4-2STs were found to be unstable. The rest were converted to this format in the following couple of years. (GW Trust)

'Convertible' 3548, originally built in 1888 as an 0-4-2ST, rebuilt in 1891 as a broad gauge 0-4-4T, 1891. (F. Moore/ MLS Collection)

'Convertible' 3554 after conversion to standard gauge 0-4-4T, c1895.
(Bob Miller/MLS Collections)

'Convertible' 3547 after conversion on a stopping train in the West Country, c1900.
(P.J.T. Reed/F.K. Davies Collection)

Chapter 3
STANDARD GAUGE LOCOMOTIVES BUILT BY THE GWR BEFORE 1923

Swindon Works was brought into use in 1843 and the first broad gauge locomotives were built there in 1846. It began to build standard gauge engines also from 1854. The first four-coupled tank engines built at Swindon were the 'Met' tanks of 1864. The Carriage & Wagon Works were opened in 1869. The standard gauge Shrewsbury and Birmingham Railway built its locomotive shed and repair shop at Wolverhampton in 1849 and the GWR took it over in 1854 and extended it as the Works with the capability of new construction as well as repairs. It was known as Wolverhampton Stafford Road and the first locomotives were built there in 1859. The first four-coupled tank engines were 2-4-0Ts built there in 1864. Sir Daniel Gooch was the GWR Locomotive Superintendent at Swindon with Joseph Armstrong at Wolverhampton. When Gooch retired in 1864, Armstrong moved to Swindon and his brother George became the Northern Locomotive Superintendent at Wolverhampton. Joseph Armstrong died in 1877 and was succeeded by William Dean. George Armstrong remained in charge at Wolverhampton until 1897. After his retirement, Dean, supported by Churchward, took over the whole locomotive department, although Dean's health was already failing. Churchward took formal command at Swindon in 1902 and new construction of locomotives at Wolverhampton ceased in 1908, the Stafford Road Works being retained as a repair facility and remained thus until taken over by the London Midland Region of BR in 1963 and the last repairs there were carried out in 1964.

Private company designed classes for the GWR
91-92, Beyer, Peacock 0-4-2ST, 1857, reb.0-4-0ST 1893

The first standard gauge GWR shunting engines were two 0-4-2 saddle tanks built in 1857 by Beyer, Peacock & Co. to their own design. They had inside frames and the saddle tank covered only the boiler barrel. No cab was provided. Their dimensions were:

Coupled wheels:	4ft 0in
Trailing wheels:	3ft 3in
Cylinders:	14in x 20in
Boiler pressure:	120lb psi
Heating surface:	646.6sqft
Grate area:	10.15sqft
Weight:	24 tons
Tractive effort:	8,330lb

The cylinders of 91 were enlarged to 15in x 20in in 1870 but it was withdrawn in 1877, its parts being used to keep 92 in service. This locomotive was rebuilt substantially at Wolverhampton in 1893, being converted to an 0-4-0 saddle tank with a cab and the following changed dimensions:

Cylinders:	15in x 20in
Heating surface:	714.8sqft
Grate area:	9.82sqft
Weight:	25 tons 19 cwt
Tank capacity:	500 gallons
Tractive effort:	11,156lb

The rebuilt 0-4-0ST No.92 at Wolverhampton Stafford Road in 1933. Note the entry to the cab is from the far side only. (Bob Miller/MLS Collections)

The rebuilt 0-4-0ST No.92 at Radyr showing the cab entry, 24 July 1938. (Bob Miller/MLS Collections)

The rebuilt 0-4-0ST No.92 shunting at Severn Tunnel Junction, March 1937. (D.B. Watkins/F.K. Davies/John Hodge Collections)

91 spent all its life in the GW's Northern Division and 92 similarly, mainly in the collieries of the Wrexham area until 1936 when it moved to South Wales. It was engaged in ballast work at Severn Tunnel Junction initially, then from 1938 at Radyr but the following year it returned to Wellington (Shropshire) mainly as a stationary engine. It was withdrawn in 1942, aged 84 and with an accumulated mileage of 785,000. It was retained a few more years purely as a stationary boiler.

342, Beyer, Peacock 0-4-0ST, 1856

An 0-4-2 saddle tank of similar design to the GW 91 and 92 was

342 at Croes Newydd, Wrexham, c1922. (Bob Miller/MLS Collections)

342 at Croes Newydd, Wrexham, in the company of 0-4-0ST No.92, c1922. (Bob Miller/ MLS Collections)

built in 1856 by Beyer, Peacock & Co. for Chester General station owned jointly by the LNWR and the Birkenhead Railway. The engine was purchased by the GWR in 1865. Its name, *Chester*, was removed and it was allotted the GW No.342. Its dimensions were mainly similar to the other Beyer, Peacock 2-4-0STs, the main changes being:

Heating surface:	572.3sqft
Grate area:	10.5sqft
Weight:	23½ tons
Coal capacity:	½ ton

It received new 15in x 20in cylinders in 1874 and was rebuilt in the same fashion as 92 in 1881 – as an 0-4-0ST. In 1897 it received a new domed boiler and 500 gallon saddle tank. Its revised dimensions were:

Cylinders:	14½in x 20in
Boiler pressure:	140lb psi
Heating surface:	655sqft
Grate area:	10sqft
Weight:	26 tons 17 cwt
Tractive effort:	10,425lb

The engine spent its long career at Chester and Croes Newydd, being withdrawn in 1931 after a mileage of 640,000.

343, Beyer, Peacock 2-4-0T, 1864

This 2-4-0 side tank was built by Beyer, Peacock in 1864 to provide a spare for two similar engines taken over from the West Midland Railway. Its dimensions were:

Coupled wheels:	5ft 0in
Leading wheels:	3ft 6in
Cylinders:	15in x 20in
Boiler pressure:	120lb psi
Heating surface:	907.74sqft
Grate area:	14sqft
Weight:	30 tons
Axleload:	11 tons
Tank capacity:	800 gallons
Tractive effort:	7,650lb

It was taken over by the GWR in 1865 and withdrawn in 1888 after achieving 510,000 miles in traffic.

Wolverhampton designed classes

45, 0-4-0ST, 1880

No.45 was built at Wolverhampton in 1880 as an 0-4-0 saddle tank to replace an earlier Shrewsbury and Birmingham Railway engine. It had inside frames and cylinders and the saddle tank covered the smokebox

45 at Croes Newydd with the later cab and safety valve cover, c1933 and on shed, 1937.
(Bob Miller/MLS Collections & W. Clark/F.K. Davies/John Hodge Collections)

and boiler barrel but not the raised firebox. Its dimensions were:

Coupled wheels:	4ft 1½in
Cylinders:	15in x 24in
Boiler pressure:	140lb psi
Heating surface:	873sqft
Grate area:	9.6sqft
Weight:	26 tons 16 cwt
Tank capacity:	480 gallons
Tractive effort:	12,980lb

The only change was the addition of an enclosed cab. It worked from Croes Newydd depot and was withdrawn in 1938 with a mileage of 430,000.

2-4-0T, 1864, reb.2-4-0ST, 1867

Twelve similar 2-4-0 side tanks were built at Wolverhampton between 1864 and 1866 with inside frames, domeless boiler, a raised firebox casing and well tanks. The well tanks were replaced by saddle tanks between 1866 and 1868. The 1864 engines were numbered 1A and 2A initially, then 17 and 18 in 1865, and 3A of 1864 and 4A of 1865 were immediately renumbered 1002 and 1003. 11 and 177 were built in 1865, the latter being renumbered 227 in 1867 and 238 from 1870. The dimensions of the 1864 locomotives were:

Coupled wheels:	5ft 0in
Leading wheels:	3ft 6in
Cylinders:	14½in x 22in
Heating surface:	713sqft
Grate area:	12sqft

346 of the '344' class of 2-4-0 side tanks built at Wolverhampton in 1865 (similar to class '17' of 1864), c1885. It was rebuilt as a saddle tank in 1867 as seen here and withdrawn in 1888. (F. Moore/Bob Miller/MLS Collections)

The later locomotives had minor variations to the heating and grate area surfaces. When rebuilt with saddle tanks, they weighed 28 tons 2 cwt and had a tank capacity of 568 gallons. Five (18, 344, 345, 1002 and 1003) were transferred to the Southern Division and later fitted with cabs. They were withdrawn between 1883 and 1893, the last survivors being 344 and 1002.

'517' class, 0-4-2T, 1868-1885

We come now to one of the first two classes of mass produced four-coupled tanks built or owned by the Great Western Railway. 156 0-4-2 side tanks of the '517' class were built at Wolverhampton between 1868 and 1885 with numerous variations and developments over the two decades. Initially they were numbered 1040-1087 and 1100-1101 but in July 1870 were renumbered 517-576. The older engines were subsequently modified to bring them in line with later developments and some modifications were still being developed up to 1915. The vast majority lasted until the late 1920s and mid-1930s, with five lasting to the end of the Second World War – 848, 1159, 1163, 1436 and 1442, the last being 1159 withdrawn just five months before nationalisation. The common features of the class were the main frame design, the arrangement of cylinders and motion, the dimensions of the standard 3 boiler and the diameter of the coupled wheels, although in the latter case the diameter grew from 5ft to 5ft 2in as a result of using thicker tyres.

The first batch of sixty engines, numbered 517-576, were built between 1868 and 1870. 517-570 were saddle tanks with the tank covering the back of the smokebox to the front of the firebox as illustrated below. 571-576 were built with side tanks. The Wolverhampton version of the GW livery was a blue-green with black boiler bands and white lining of the tank panels until 1894 when the Swindon standard green was applied.

The dimensions of these sixty locomotives were:

Coupled wheels:	5ft 0in
Trailing wheels:	3ft 6in
Cylinders:	15in x 24in
Boiler pressure:	140lb psi
Heating surface:	808sqft

The first example of the class, 517, built in 1868 with saddle tank and converted to a side tank engine in 1880. This photograph was therefore taken in the 1870s before any modification. (F. Moore/MLS Collection)

567 was also built as a saddle tank and appears to have remained so, being reboilered in 1888 and was sold to the Bishop's Castle Railway in 1905 where it became the railway's No.1 and was not withdrawn until 1936. It is seen here probably in the 1880s as built apart from the cab additions including a rear weatherboard. (F. Moore/Bob Miller/MLS Collections)

556 as built in 1869, rebuilt as a side tank in 1890 and withdrawn in 1933. (F.K. Davies/John Hodge Collections)

Grate area: 12.33sqft
Weight: 27 tons 2 cwt
Axleload: 9 tons 18 cwt
Tank capacity: 620 gallons
Coal capacity: ¾ ton
Tractive effort: 10,710lb

Later, after various minor modifications, the side tank engines were weighed at slightly over 32 tons. Then a second batch of side tanks similar to 571-576 were constructed continuously from 1873 to 1878. The first twenty-four were numbered 826-849, the next dozen, 1154-1165, the next twelve, 202-205 and 215-222 and the final batch 1421-1444. The 826 and 1154 series differed only in the increase in heating surface to 914sqft, the 202 series in a reduction of the tank capacity to 600 gallons. From 1421 the appearance changed slightly, with brass covered domes and copper topped chimneys. 1421 and a few of that series were fitted with condensing gear for heating the feed water and had pumps rather than injectors.

The final batch, 1465-1488, were constructed between January 1883 and November 1885 and the major change was an increase in cylinder diameter to 16in. The number and size of boiler tubes varied, reducing the total heating surface to 779sqft. The last six, 1483-1488, had a longer wheelbase and outside axleboxes on the trailing wheels and an enlarged heating surface of 914sqft. The increase in cylinder size boosted the tractive effort to 11,986lb and all the class had 16in cylinders by the end of the 1890s. 1443 had a further enlargement to 16½in and this gradually became standard for

1164 of the second batch of the '517' class built in March 1876. This series was fitted with spectacle plates before and aft. 1164 is seen at Wolverhampton Stafford Road, c 1880. This was the last series with individual brass numbers on the tank side rather than what became the standard GW numberplate. The engine's holly green and pea green and white tank lining has been 'guivered', a form of special cleaning 'decoration' favoured and insisted on by some drivers of the period.
(F. Moore/MLS Collection)

204 of May 1876 with the GW standard numberplate, c1880.
(F. Moore/MLS Collection)

the class. 1427 was even fitted with 17in cylinders in 1908 but that remained a 'one-off'.

The saddle tanks were nearly all converted to side tank engines between 1876 and 1886 and basic open cabs were fitted in the same period to replace the basic weatherboards of the 1870s. The frames were extended on nine engines between 1891 and 1895 to increase the bunker and cab space and both Wolverhampton and Swindon Works undertook this modification to the class between 1894 and 1915. 1477's tanks were increased to give 900 gallons capacity, and this became standard for the class from 1900. Many engines received enclosed cabs during the Churchward era and from 1924 a Collett new style

1421 of April 1877 with brass dome and condensing gear on Stafford Road shed, c1880. (F. Moore/ MLS Collection)

of enclosed cab and his standard bunker became the norm.

Twenty-nine of the '517' class were auto-fitted from 1904 and were painted a 'coffee brown' to match the trailer. Two locomotives, 533 and 833, were disguised as railmotors, the coaching outline extended over the engines in 1906 until removed in 1911. After the First World War, a number of these auto-fitted engines were painted in the 'crimson lake' livery to match the Churchward auto-trailer colours. The auto engines reverted to the standard GWR mid-chrome green livery after 1923.

The R3 boiler was replaced by the S boiler as standard from 1890 onwards and 1477 received a 'U' Belpaire boiler in 1894. Ten others received this type of boiler between 1900 and 1902. 831 and 1466 were fitted with extended smokeboxes in 1914/5 with the intention of equipping them with a superheater but this was not followed through.

834, built in 1874, fitted with an open cab in the 1880s and not withdrawn until 1933, seen in the 1890s still in the Wolverhampton style livery. Note it still has the Wolverhampton style bunker with the flared top ridges. (Loco Publishing Co./Bob Miller/ MLS Collections)

Above left: **530 of** 1868 looking spick and span on Stafford Road shed with an enclosed cab provided in the first decade of the twentieth century. (MLS Collection)

Above right: **1421, which** was one of the nine 1891-5 engines rebuilt with extended frame, new 'U' Belpaire boiler, seen here immediately after the rebuilding, c1900. (Loco Publishing Co./Bob Miller/MLS Collections)

Below: **551 of** 1869, rebuilt as a side tank shortly after 1876, reboilered in 1895 and in GW standard green livery and Swindon style shallow bunker with straight back. Note the outside axlebox on the trailing wheels and ramp on the running plate. Seen ex-works at Swindon, c1910. (F. Moore/MLS Collection)

565 of similar vintage but with Belpaire boiler fitted in 1911, open cab and Swindon style bunker, c1911. (F. Moore/MLS Collection)

Below left: **519 of** 1868 which was one of the nine 1891-5 engines rebuilt with extended frame, new Belpaire boiler and top feed, increased bunker and cab space, c1920. (Bob Miller/MLS Collections)

Below right: **The prototype,** 517, in GWR times now with half cab and outside rear axleboxes, but retaining the Wolverhampton style bunker. 517 received five different boilers in its career and was withdrawn in 1934. (Bob Miller/MLS Collections)

Standard Gauge Locomotives built by the GWR before 1923 • 33

205 with Belpaire boiler, outside rear axlebox, enclosed cab and a Collett style bunker, near the end of its life, c1933. (MLS Collection)

1154 of 1875 in its final form similar to 205 above but seen from the other side, outside Stafford Road Erecting Shop, c1930. (Photomatic/MLS Collection)

34 • FOUR-COUPLED TANK LOCOMOTIVE CLASSES BUILT BY THE GREAT WESTERN RAILWAY

Above left: 1466 with extended smokebox fitted in 1914, Churchward half cab, Swindon style bunker and outside rear axlebox, c1920. (Loco Publishing Co./ Bob Miller/MLS Collections)

Above right: 1472 of the final 1883 series as built apart from the later addition of the Collett bunker and outside trailing wheel axlebox, c1930. (Bob Miller/ MLS Collections)

1438 with Belpaire boiler, enclosed cab and GW Collett bunker at Staines, 3 July 1930. (Bob Miller/MLS Collection)

Operation

In the nineteenth century the Wolverhampton standard gauge 0-4-2Ts operated mainly in the Northern Division whilst similar work in the Southern Division was handled by the Swindon built 2-4-0 'Met' tanks. Their initial work was on local and suburban passenger services, particularly around Birmingham, Wolverhampton, Chester, Shrewsbury and Worcester. A few were in Central Wales working from the Wrexham and Shrewsbury areas and also in the Newport Division and a few in the Neath area operating on the GW lines. At the turn of the century, the Dean/Churchward 2-4-2Ts of the 36XX class and Churchward's 2-6-2Ts gradually became available and took over much of this suburban work and the '517s' were relegated to branch line passenger work, local goods traffic and shunting throughout the GW system.

In 1904, with their release from the Birmingham area, they became available for branch line auto-train work and some thirty locomotives were fitted with auto-train equipment during the first decade of the twentieth century. One locomotive became well known – 1473 after its royal train duties on the Woodstock branch for the Prince of Wales (the future Edward VII) in 1896 – and it continued to work the Oxford-Woodstock branch until its withdrawal in 1935, retaining its name of *Fair Rosamund* that was applied for that occasion.

Some worked in the latter days in the London area particularly around Southall/Uxbridge and Ealing. The last examples were at Swindon (1161, 1436 and 1442) withdrawn in 1944/5, 1163 at Weymouth (withdrawn in 1946) and the final one, 1159 from Oxford in August 1947. The following photographs show the diversity of distribution.

1473 of the final 1883 series named *Fair Rosamund* after a Blenheim-based mistress of King Henry VIII, in preparation for a royal visit to Woodstock and Blenheim in 1896, seen here at Oxford, c1930. (MLS Collection)

It is difficult to find moving train photographs of the first batch of saddle tank 517s as nearly all were rebuilt with side tanks by the early 1880s. However, 569 was one of five saddle tanks not converted until the mid-1890s and it was captured hauling a local mixed train around 1890. (F, Moore/Bob Miller/MLS Collections)

538, after conversion to side tank format in the early 1880s, operates a substantial local passenger train at an unknown location, c1900. (F. Moore/Bob Miller/MLS Collections)

202 of 1876 on arrival at Cardigan with the branch train from Whitland, c1910. (Bob Miller/MLS Collections)

1165 with an auto train trailer at Trumpers Crossing, near Osterley Park in West London, c1905. This was shortly after auto train working was introduced. Note the spelling of the 'Halte'. (F.K. Davies/John Hodge Collections)

In 1906 two '517s', 533 and 833, were disguised as railmotors. Here is 833 working the auto-train at Trumpers Crossing Halte, c1906. The carriage disguise was removed in 1911. (F.K. Davies/John Hodge Collections)

Auto fitted 518 and auto coach 86, both painted crimson lake, thought to be the Slough-Windsor set at its introduction in 1912. (Loco Publishing Co./Bob Miller/MLS Collections)

Standard Gauge Locomotives built by the GWR before 1923 • 39

Above left: **832 after** rebuilding with cab, Swindon style bunker and outside rear axleboxes on a local goods service at an unknown location, c1920. (Bob Miller/MLS Collections)

Above right: **837 with** another '517' and a Dean 4-4-0 at Birmingham Snow Hill station, c 1900. Note 837 has wing plates from the smokebox to the running plate. (Bob Miller/MLS Collections)

Below: **1155 of** 1875, rebuilt with the Swindon high-sided flat bunker and enclosed cab, with the Dinas Mawddwy branch train, c1935. (LGRP/MLS Collection)

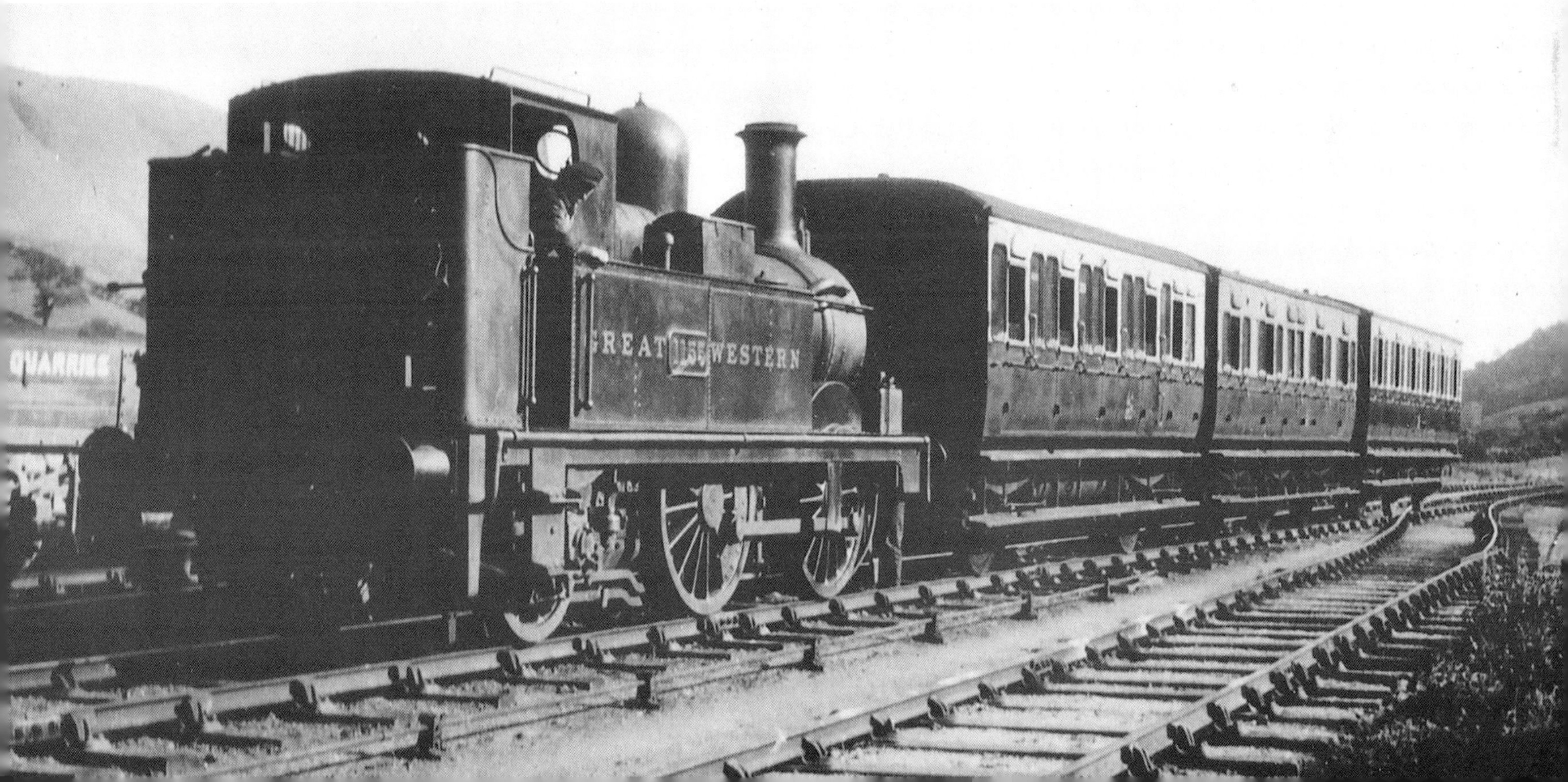

Above left: 204 with an R4 boiler (rebuilt between 1901 and 1904) with a local passenger train at Bristol Temple Meads, c 1912. (W. Vaughan-Jenkins/F.K. Davies/John Hodge Collections)

Above right: 1466 heads a local passenger train at an unknown location, before its smokebox was extended in 1914. (Bob Miller/MLS Collections)

574 of 1870 arrives at Oswestry with a local goods train from the Cambrian lines, 28 August 1926. (H.C. Casserley/Bob Miller/MLS Collections)

833 at Southall with trailer car No.88, March 1920. (J.N. Maskelyn/F.K. Davies/John Hodge Collections)

Below left: **205 at** Exeter St David's, May 1927. (F.H.C. Casbourn/F.K. Davies/John Hodge Collections)

Below right: **547 at** Staines on an auto-sandwich train, 1926. (F.M. Gates/F.K. Davies/John Hodge Collections)

42 • FOUR-COUPLED TANK LOCOMOTIVE CLASSES BUILT BY THE GREAT WESTERN RAILWAY

Above left: **530 at** the head of a suburban train at an unknown location, c1920. (W.H. Whitworth/F.K. Davies/John Hodge Collections)

Above right: **555 comes** off the line from Stratford-on-Avon at Hatton with a stopping train for Leamington Spa, c1928. (W.L. Good/F.K. Davies/John Hodge Collections)

830 on arrival at Blenheim & Woodstock with the 4pm from Oxford, 13 June 1931. (A.W. Croughton/F.K. Davies/John Hodge Collections)

Standard Gauge Locomotives built by the GWR before 1923 • 43

Above left: **An interesting** formation of 1157 with the Wrexham auto-train at Saltney Junction, 7 September 1932. (R.E. Thomas/F.K. Davies/John Hodge Collections)

Above right: **1428 at** an unidentified location, c1930. (G.R. Grigs/F.K. Davies/John Hodge Collections)

1465 and auto trailer 155 at Yeovil, 21 May 1935. (H.C. Casserley/F.K. Davies/John Hodge Collections)

3571-3580, 0-4-2T, 1895

Ten 0-4-2 side tanks were built between 1895 and 1897 as a development of the '517' class (after 1477 was tested with this boiler version in 1894) and were designated as '3571s'. They were numbered 3571-3580 and differed from the 1883 batch of '517s' only in the extension of the frame at the rear of the driving wheels which lay outside the trailing wheels with only the axlebox outside the frame. Six (3571, 3573, 3574, 3576, 3577 and 3580) had GW Collett bunkers from the mid-1920s and 3576 had an enclosed cab. They were equipped with the 'U' Belpaire boiler.

Some were fitted with extended smokeboxes. Their dimensions were:

Coupled wheels:	5ft 2in
Trailing wheels:	3ft 8in
Cylinders:	16½in x 24in
Boiler pressure:	140lb psi
Heating surface:	1,018.75sqft
Grate area:	15sqft
Weight:	40¼ tons
Axleload:	14¼ tons
Tank capacity:	900 gallons
Tractive effort:	12,541lb

They survived well into the post-Grouping period, only two being withdrawn before the Second World War (3572 in 1928 and 3576 in 1929). Two more were withdrawn during the war, but three – 3574, 3575 and 3577 – went into BR ownership in 1948 and all lasted a further year being withdrawn in 1949, the last, 3574, in December.

The only one of the class to be fitted with an enclosed cab, 3576, seen at Chester in 1927.
(J.E.N. Ashworth/F.K. Davies/John Hodge Collections)

Standard Gauge Locomotives built by the GWR before 1923 • 45

3575 at Chester, 2 January 1938. The additional outside framing at the rear of the engine is very clear in this photograph. The engine has a Wolverhampton style bunker with spectacle plate.
(MLS Collection)

3579 with an extended smokebox and Wolverhampton style bunker with double beaded fender at Chester shed, 1 January 1939.
(D. Darby/MLS Collection)

46 • FOUR-COUPLED TANK LOCOMOTIVE CLASSES BUILT BY THE GREAT WESTERN RAILWAY

Above left: **3571 on** a Chester-Birkenhead train at Hooton, 1930. (W.P. Riley/Bob Miller/MLS Collections)

Above right: **3578 at** Hooton also on a Chester-Birkenhead train, c1932. (J.A. Peden/MLS Collection)

Operation
Initially, they were retained in the Northern Division around Chester and Birkenhead hauling branch and local passenger services until displaced by the 36XX 2-4-2Ts. A couple (3576 and 3579) were seen working from Pontypool Road in the first decade of the new century and 3576 worked from Bristol for ten years or so spanning the First World War. They were mainly at Chester in the 1920s, though 3573 and 3574 went to Worcester in 1929, while 3571 and 3580 were allocated to Swindon in 1935. 3577 was suddenly allocated to Carmarthen in 1939 and 3575 also went to South Wales, but both were used on shed for stationary steam raising. 3572 was withdrawn in 1928 after reaching 666,505 miles and the last survivor, 3574, ran over a million (1,101,474).

Swindon designed classes
117, 4-4-0T, 1854
This locomotive appears to be a mystery as Swindon were not building standard gauge engines in 1854. It seems that it may have been built for a contractor, a Mr. Marshall, and was a 4-4-0T. It is possible that it was rebuilt from a broad gauge engine of the same wheel arrangement such as the 'Corsair' class built in 1849. Mr Marshall was bankrupted in 1861 and the fate of the engine after that is uncertain, though it is probable that it was broken up in 1863. Its known dimensions were:

Coupled wheels:	4ft 8½in
Bogie wheels:	3ft 0in
Cylinders:	15in x 20in
Weight:	27 tons 14 cwt

320-1, 2-4-0WT, 1864
Another couple of standard gauge engines built by Swindon as early

Drawing of the GW standard gauge 'Met Tank' of 1864, before rebuilding as a 2-4-0 tender engine. (E.L. Ahrons/RCTS)

as 1864. These were two standard gauge 'Met Tanks' fitted with well tanks and condensing gear for working over the underground lines of the Metropolitan Railway. They had steeply inclined cylinders like the Beyer, Peacock engines built for the 'Met' and inside frames. The known dimensions of the two engines were:

Coupled wheels:	5ft 6in
Leading wheels:	3ft 7in
Cylinders:	15in x 24in
Heating surface:	817.15sqft
Grate area:	12.57sqft
Weight:	30¼ tons
Axleload:	11 tons

The condensing gear gave much trouble (it was the first application on the standard gauge) and the engines were rebuilt in 1867 and 1873 respectively as tender engines and the condensing gear was removed. They were later banished to the Hereford area and were withdrawn in 1881, having run less than 200,000 miles each, nearly all in tender form.

The '455' class 'Metro Tank' 2-4-0T, 1869-1899

The standard gauge 2-4-0 side tanks, the 'Metropolitan Tanks', were so named as they were built by Swindon over the latter part of the nineteenth century to have general route availability including over London's Metropolitan Railway. Built in three basic sizes, the 'Small' were from lot 18 of 1869, the 'Medium', starting with lot 25 in 1871 and concluding with lot 96 in 1894, and the 'Large' – lots 117 and 119 of 1899. The engines from lot 18 were numbered 455-470 and 3-6, and the series was officially known as the '455' class. These engines had condensing gear. They soon became better known as the 'Metro Tanks' or even the abbreviated 'Met Tanks'. They replaced the twenty-two broad gauge 'Metropolitan Tanks' of 1862. Their dimensions were:

Coupled wheels:	5ft 0in
Leading wheels:	3ft 6in
Cylinders:	16in x 24in
Boiler pressure:	140lb psi
Heating surface:	1,080sqft
Grate area:	15.75sqft
Weight:	33 tons 4 cwt
Axleload:	12 tons 2 cwt
Wheelbase:	15ft 3in
Tank capacity:	740 gallons
Tractive effort:	12,186lb

These 'small' Metro tanks were reboilered and altered with a straight running plate extending the wheelbase by 6 inches between the leading pair of wheels and the first coupled axle in the 1880s or 1890s. They received enlarged tanks of 820 or 860 gallons capacity towards the end of the century. They were fitted with ATC between 1929 and 1931. Four of the class were withdrawn as early as 1906 (including the prototype 455) but others remained in service until the early 1930s, the last of this series of 'small' Metro tanks being 464 withdrawn in December 1934.

Two years later, Swindon lot 25 followed, with another twenty locomotives numbered 613-632. They were enlarged versions of the 455 series with longer wheelbase. 613-622 were fitted with condensing apparatus. Their changed dimensions were:

Heating surface:	1,208sqft
Weight:	33 tons 8 cwt
Tank capacity:	800 gallons

The Metro Tank No. 457 as built in 1869 with condensing gear and very basic crew protection. It is seen in the original livery and lining and was probably taken in the 1880s. (Loco Publishing Co./Bob Miller/MLS Collections)

The prototype, 455, after modification to straight running plate, wheelbase extension and provision of a larger tank and open cab, seen around 1900 and before its early withdrawal in 1906. (Bob Miller/MLS Collections)

No.5 of the 1869 series after frame extension, tank enlargement and cab provision, c1920. This locomotive was not withdrawn until 1932. (Bob Miller/MLS Collections)

Standard Gauge Locomotives built by the GWR before 1923 • 49

No. 457 again, the photograph taken some forty years after the previous one of this engine, after overhaul at Swindon in 1923 with new B4 boiler and enclosed cab and painted green with Great Western lettering and numberplate moved back to the cab side. (F. Moore/Bob Miller/MLS Collections)

No. 632 of the 1871 series seen later after provision of larger tank and cab. Note that the running plate is straight and front wheel axleboxes are outside the frame. These changes were made in 1891 and the engine was withdrawn in 1929. This shot was taken c1905. (F. Moore/MLS Collection)

No. 626 in later years with enclosed cab. It was withdrawn in December 1932. (Bob Miller/MLS Collection)

Again, a few were withdrawn in the first decade of the twentieth century, 623 withdrawn in 1903 being the first condemnation from the whole class.

Another twenty of the same 'medium' size followed in 1874 numbered 967-986, twenty more in 1878, 1401-1420, another twenty in 1881, 1445-1464. These were of broadly the same dimensions as 613-632, with minor variations only in heating surface although the weight appears to have increased to 34 tons, then 36¾ tons for the 1878 engines. These were built with straight running plates and the axles of the leading pair of wheels outside the frame. Like the 455 series, they received the larger tanks, each batch being slightly different. Some had 860 gallon capacity. 1401-1420 were fitted with even larger 1,080 gallon tanks around 1900. The first ten of each new series were fitted with condensing gear, thus 967-976 and 1401-1410. Some fifty in all were required for the London area with the potential of working over Metropolitan lines. The non-condensing engines had open cabs from 1878 onwards. Later in the 1920s several received enclosed cabs and GW style Collett bunkers.

There was then a gap of ten years before another batch, lot 91, of the 'medium' Metro Tanks, was built in 1894. It consisted of ten engines numbered 1491-1500. These had an increased heating surface of 1,308sqft and grate area of 16.28sqft. They were heavier at 39 tons 2 cwt. Two years later, a further batch was built, lot 95, numbered 3561-3570, heavier still at nearly 41 tons. Most of these engines lasted until the late 1930s with seven lasting until the 1940s and two, 3561 and 3562, not being withdrawn until the BR era in 1949.

Condensing tank No.974 as built in 1874 and before reboilering in 1888. (F. Moore/MLS Collection)

Standard Gauge Locomotives built by the GWR before 1923 • 51

967 also built with condensing apparatus in 1874. It is seen here probably in the late 1880s. (Bob Miller/ MLS Collections)

967 again after reboilering and overhaul either in 1901 or 1904, probably at Stafford Road Works. (Bob Miller/ MLS Collections)

1406 built in 1878 and fitted with 1,080 gallon tanks in 1898. Its condensing gear was removed after electrification of the Hammersmith & City line in 1906. It is seen here at Old Oak Common, c1920. It was withdrawn in 1929.
(Bob Miller/MLS Collections)

1450 built in 1881 with Dean cab and bunker, seen c1920. It was withdrawn in 1928.
(Bob Miller/MLS Collections)

1453 built in 1882 and seen as rebuilt with Collett enclosed cab and bunker, ex-works at Swindon, in the late 1920s. It was not withdrawn until 1933.
(Bob Miller/MLS Collections)

1494 as built in 1892 with Dean cab and bunker, c1920. It was withdrawn in 1937.
(MLS Collection)

3568 built in 1894 with condensing gear, and open cab, and later fitted with 1,080 gallon tanks, c1905. (Bob Miller/MLS Collection)

3561 of 1894 lasted into BR days and is seen in the last year of its life, on 10 June 1949. It has been stripped of its condensing gear but has acquired large tank, Collett cab and bunker. (J.D. Darby/MLS Collection)

The final two lots, 117 and 119 of 1899, and numbered 3581-3600, were the twenty 'large' Metro Tanks which were built with 1,100 gallon tank capacity and weighed around 42 tons. The axleload was raised to 14¼ tons. These were fitted with condensing gear until the electrification of the Metropolitan and were fitted later with Collett cab and bunker. 3593 was extended to become a 2-4-2T in 1905 with 3ft 8in trailing wheels under a larger bunker and higher enclosed cab roof. Its wheelbase was lengthened to 23ft. 3596 also received similar cab and bunker form in 1908 though it remained a 2-4-0T. 3593 weighed 48 tons 6 cwt. Later boilers were pressed at 150 or 165lb psi raising the tractive effort to 12,635 and 13,895lb respectively and some received Belpaire fireboxes. The experiment on 3593 does not seem to have been successful and it was the first withdrawal from this series in 1927. 3600 was renumbered 3500 in 1912 to avoid confusion with the 36XX class. Just two were withdrawn before the Second World War and eight lasted into the BR period, 3599, 3586 and 3588 being withdrawn in October, November and December 1949 respectively.

Standard Gauge Locomotives built by the GWR before 1923 • 55

3583, a 'large' Met Tank as built in 1899, at Paddington, c1907.
(F.K. Davies/John Hodge Collections)

3583, after rebuilding with Collett enclosed cab and later bunker, at Paddington, c1935. 3583 was withdrawn in 1947.
(LPC/F.K. Davies/John Hodge Collections)

Above left: **3593 as** rebuilt in 1906 as a 2-4-2T with high roof and extended frame, c1910. (LGRP/F.K. Davies/John Hodge Collections)

Above right: **The 1906** rebuilt 3593 as a 2-4-2T with high vaulted roof, large side tanks and extended bunker, c1925. (W. Potter/MLS Collection)

3595 ex-works in the 1930s with Collett enclosed can and bunker. It was withdrawn at the beginning of 1945.
(Bob Miller/MLS Collections)

Standard Gauge Locomotives built by the GWR before 1923 • 57

Above left, above right and below: **Three views** of 3596 rebuilt with high vaulted cab roof in 1908 but retaining 2-4-0T wheel arrangement. It is seen first at Slough in 1919, then much later with auto coach and on shed at Southall, at the end of the Second World War and just before withdrawal in March 1945. ((1) J.N. Maskelyn/F.K. Davies/John Hodge Collections (2) Colling Turner/MLS Collection (3)MLS Collection)

Operation

Usually around fifty of the total of 140 engines of the class were stationed in the London area and fitted with condensing gear for operation over the Metropolitan lines. As new batches were built, they tended to displace some of the older engines in the area. Outside the London District, which included Reading and Oxford, they were until around 1905 dispersed over most of the Southern Division, similar work in the north being the realm of the 0-4-2T '517' class. There were concentrations at Bristol and Gloucester. They could display a turn of speed on Paddington-Oxford semi-fasts or on services like Gloucester-Cardiff or Bristol-Weston-super-Mare-Taunton.

When the Churchward 2-6-2Ts began to arrive in numbers in the first decade of the twentieth century, some made their way north to Worcester, Wolverhampton, Wellington, Croes Newydd and Chester, but a significant number, especially nearly all the 'large' tanks in the 3581-3600 series, continued to perform sterling work on the Paddington suburban services until the arrival of the 61XX 2-6-2Ts in the early 1930s and the fitting of condensing gear to 57XX pannier tanks 9700-9710. A few gained access to South Wales among the myriad former engines of the absorbed companies there although, apart from Llanelly, few went to the far west. Just one, 617, was said to be the only example of the class to be allocated to a former Cambrian depot – Aberystwyth. Forty of the class, including ten of the 1899 'large' Metro Tanks, were fitted with auto apparatus, the first (3581 and 3582) in 1925, the majority between

The 1874 built 985 in original condition and with condensing gear on a main line stopping train, probably in the Reading area, c 1890.
(Bob Miller/MLS Collections)

1928 and 1930 and the last (3586) in 1934. Those fitted with push-pull equipment in the London District would include the auto services to Uxbridge and the Greenford-Ealing branch as well as branch services in the Oxford area. The lone 2-4-2T, 3593, after initially working in Cornwall from Carn Brea and St Blazey, was based at Reading for many years until its withdrawal in 1927.

All the 1869-78 engines had been withdrawn by 1934 and the 1881 examples before the start of the Second World War, but a few of the 1892 and 1894 engines survived the war and the vast majority of the 1899 'large' tanks were not withdrawn until the mid/late 1940s. The last survivors, month of withdrawal in 1949 and their last base were:

3561 (auto)	at Swindon	October
3562	at Oxford	February
3582 (auto)	at St Blazey	November
3586 (auto)	at Llantrisant	November
3592	at Carmarthen	April
3599 (auto)	at Radyr	October

The photographs of the 'Metro Tanks' on train services illustrate some of the widespread locations at which they operated for forty or fifty years.

The 1869 built No.3 departing from Penrhyn with the Falmouth branch train, c1890. (F.K. Davies/John Hodge Collections)

Above left: **1461, built** in 1882, at Cheltenham St James, c1910. (Bob Miller/MLS Collections)

Above right: **627 'built** in 1871' at an unknown location in the first decade of the twentieth century. (Bob Miller/MLS Collections)

Below: **1411, built** in 1878, at Old Oak Common West with a down commuter train, c1910. (Loco Publishing Co/Bob Miller/MLS Collections)

Standard Gauge Locomotives built by the GWR before 1923 • 61

Above left: 'Large' Metro Tank 3587 and 1894 built 3565 depart from Paddington with a heavy evening commuter train in 1911. A class '850' saddle tank used for Paddington-Old Oak Common ECS duties is glimpsed in the background. (Bob Miller/MLS Collections)

Above right: No.6 of the original lot 18 of 1869 approaching Paddington with a suburban train, c1905. (Bob Miller/MLS Collection)

1894 built 3565 again, but many years later in the mid-1920s, departs from Paddington with a commuter train. It is in its final form with Collett enclosed cab and bunker. (Bob Miller/MLS Collections)

Above left: **1878 built** 1407 shunting at Slough, 17 May 1930. (H.C. Casserley/Bob Miller/MLS Collections)

Above right: **Large tanked** 3570 of 1894, still with condensing gear and with only spectacle plates to protect the crew, awaits its next turn of duty outside Paddington station, c1930. (MLS Collection)

632 of 1871 awaits departure from Bristol Temple Meads with a train for Avonmouth or Bath, c1910. (LGRP/Bob Miller/MLS Collections)

Standard Gauge Locomotives built by the GWR before 1923 • 63

Above left: Sixty-one-year-old No.5 of 1871 on the branch train for Taunton at Chard station, 2 August 1930. (H.C. Casserley/Bob Miller/MLS Collections)

Above right: 980 'built in 1871' stands ready for service at Worcester Shrub Hill, 7 April 1923. (A.W.Croughton/F.K.Davies/John Hodge Collections)

Below: 975 at Newton Abbot with the auto train for Bovey Tracey, c1925. (A.P. Murray/F.K. Davies/John Hodge Collections)

1498 'built in 1892' awaits duty at Oxford, 13 March 1926.
(A.W. Croughton/F.K. Davies/ John Hodge Collections)

1894 built 3569 with large tanks pilots Dean/ Churchward 2-4-2T 3601 on a heavy evening commuter train departing from Paddington, May 1927.
(J.N. Maskelyn/F.K. Davies/ John Hodge Collections)

Standard Gauge Locomotives built by the GWR before 1923 • 65

1499 at Truro with a local branch train, 31 July 1933. 1499 survived the Second World War, not being withdrawn until 1946. (P.J.T. Reed/F.K. Davies/ John Hodge Collections)

The unique 2-4-0T 455 class with the high vaulted roof modified in 1908, 3596, with a semi-fast down suburban train, first stop Hayes, between Kensal Green and Old Oak Common, passing 2038 with a down freight, January 1933. (J.M. Craig/F.K. Davies/ John Hodge Collections)

A much-rebuilt 1869 engine, 457, standing on Oxford shed with the depot breakdown train, 16 September 1933. 457 was withdrawn the following year aged 65 years. (F.M. Butterfield/ F.K. Davies/John Hodge Collections)

1899 built 'large' Met Tank, 3585, enters Oxford station with auto trailer 110 and the branch train from Witney, January 1947. 3585 was withdrawn exactly a year later having just survived to become an asset of the nationalised railway. (H.W. Robinson/F.K. Davies/ John Hodge Collections)

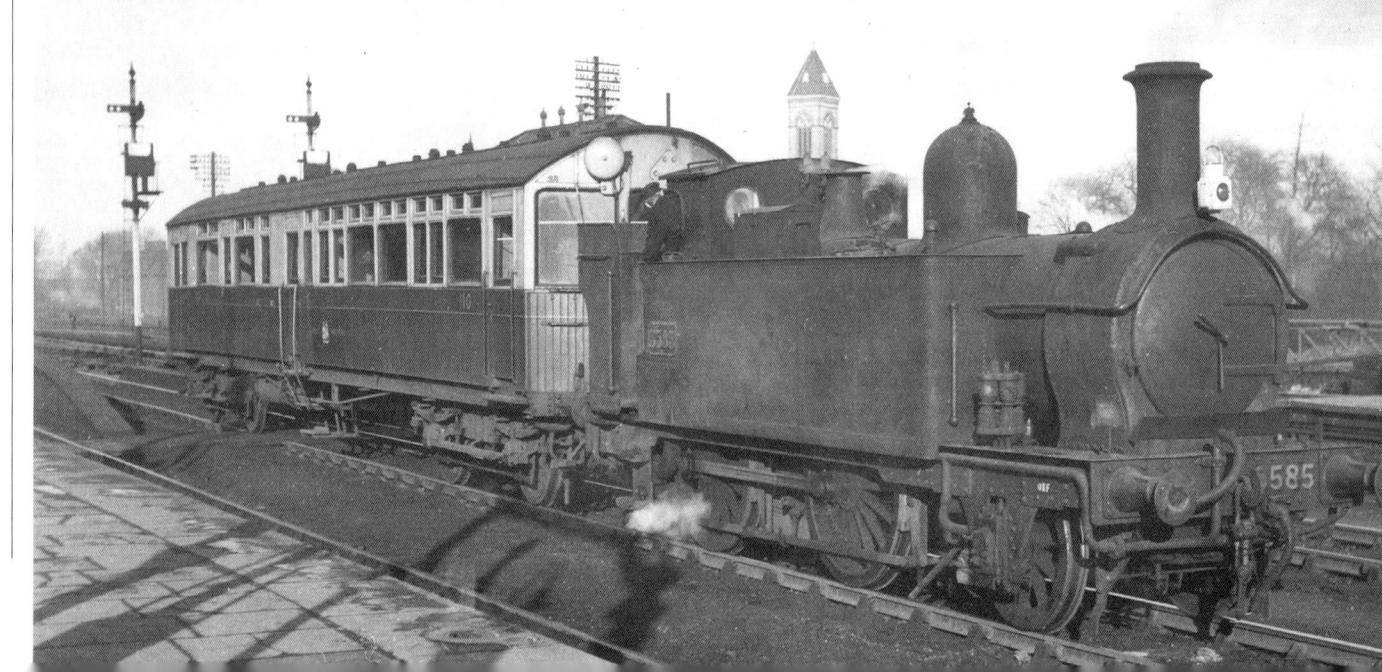

3511-3520, 2-4-0T, 1885

Dean designed and had constructed at Swindon ten double-framed 2-4-0 side tanks in 1885 which were basically tank versions of his 2-4-0 'Stella' class. With larger cylinders, they were more powerful than the Metro Tanks, but no more were built and less than ten years later in 1894 and 1895 they were converted to tender engines of the 'Stella' class. A very similar group of ten were built for work on the broad gauge (see page 18). The dimensions of the standard gauge engines were:

Coupled wheels:	5ft 1in
Leading wheels:	3ft 7in
Cylinders:	17in x 26in
Boiler pressure:	140lb psi
Heating surface:	1,209.86sqft
Grate area:	15.2sqft
Weight:	43 tons 3 cwt
Axleload:	16 tons 16 cwt
Tank capacity:	1,000 gallons
Tractive effort:	14,658lb

They were initially employed on banking and other duties through the Severn Tunnel (for which purpose they were at first given condensing apparatus), but most were transferred shortly afterwards to the London area where they worked trains to Oxford via Reading and High Wycombe/Thame. 3512, 3518 and 3520 remained in the west, probably around Bristol on services to Cardiff or Weymouth.

3516 as built in 1885 with condensing gear for operation through the Severn Tunnel. It is seen here, probably at Severn Tunnel Junction shortly after delivery at the end of 1885. Note the 'guivering' decoration of the double frame.
(LPC/F.K. Davies/John Hodge Collections)

3512 after the removal of the condensing gear. It is seen here c1890.
(LPC/F.K. Davies/John Hodge Collections)

3521-3540, 0-4-2T, 1887, reb 0-4-4T, 1891

After the 1885 2-4-0 double-framed tanks, Dean built forty 0-4-2 side tanks in 1887, the first twenty of which, numbered 3521-3540, were standard gauge and the last twenty, 3541-3560, were broad gauge 'convertibles' which were described earlier (see pages 18-19). They had similar boilers to the Dean Goods 0-6-0s with much higher pressure than the previous four-coupled tanks, however it was reduced to 160lb psi later. Their dimensions were:

Coupled wheels:	5ft 0in
Trailing wheels:	4ft 0in
Cylinders:	17in x 24in
Boiler pressure:	180lb psi (160lb psi later)
Heating surface:	1,194.6sqft
Grate area:	18.9sqft
Weight:	47 tons 18 cwt
Axleload:	16 tons
Tank capacity:	1,019 gallons (3524-40, 920 gallons)
Tractive effort:	17,687lb

There was initial complaint about unsteadiness in riding and some modifications were quickly made to the tanks and trailing wheel arrangements. Even more concerns were expressed with the riding of the 'Convertibles' which had saddle tanks and the last one, 3560, was built as an 0-4-4 side tank. The replacement of the trailing pair of wheels by a rear bogie improved the riding to such an extent that the whole class of both standard gauge engines were rebuilt as 0-4-4Ts in 1890/1, after the Convertibles had been rebuilt first. The bogie wheels were reduced in diameter to 3ft 6in, the new side tanks increased the capacity to 1,033 gallons and the weight increased to 48 tons, although the axleload on the coupled wheels fell to 15¼ tons. They were further subject to a complete rebuilding between 1899 and 1902 as 4-4-0s as described in my earlier book on the GW's small-wheeled 4-4-0 classes (see the bibliography).

The standard gauge engines worked mainly in the London and Bristol districts both before and after rebuilding as 0-4-4Ts, whereas the Convertibles operated on the broad gauge west of Exeter until 1892, after which both series were interchangeable and many of the former standard gauge engines gravitated to the west. However, despite their rebuilding, they still seemed to court problems and 3521, piloting a regauged 3548, derailed at speed on the curves near Bodmin Road in Cornwall.

3537 still in the initial rigid 0-4-2 wheel base as built in 1887, before modification and rebuilding as an 0-4-4T. (F. Moore/MLS Collection)

3527 after rebuilding as an 0-4-4T at Plymouth North Road. This photo is said to be the first train hauled by a rebuilt engine of the class on the new standard gauge, and only a month after 3527's rebuilding, May 1892. (P.J.T. Reed/F.K. Davies/John Hodge Collections)

Although the track was blamed, it seems that the track deterioration was caused by the stresses from this class of rigid wheelbase tank locomotives that handled much of the traffic west of Plymouth at the time. A couple of the former Convertibles, 3543 and 3556, had mishaps also but in the Newport and Worcester districts and these undoubtedly led to the decision to rebuild the engines with tenders at the end of the century.

3600-3630, 2-4-2T, 1900

Dean, assisted by Churchward, designed further four-coupled tank engines in 1900, though a rear axle was added to support an enlarged bunker to create a 2-4-2 side tank given the number 11. After trials which proved satisfactory, further extensions to the 3571 0-4-2Ts or the 3501 2-4-0Ts were set aside and were replaced by twenty of the new 2-4-2Ts with numbers 3601-3620. The first of these was put into traffic in February 1902 and Churchward built a further ten, 3621-3630, in 1903. After that he turned his attention to a larger 2-6-2T and no more engines of this wheelbase were built for the GWR. Their dimensions were:

	11	3601-3620	3621 – 3630
Coupled wheels:	5ft 2in	5ft 2in	5ft 2in
Front & trailing wheels:	3ft 8in	3ft 8in	3ft 8in
Cylinders:	17in x 24in	17in x 24in	17in x 24in
Boiler pressure:	180lb psi	180lb psi	180lb psi
Heating surface:	1,561.65sqft	1,561.65sqft	1,425.68sqft
Grate area:	21.35sqft	21.35sqft	20.35sqft
Weight:	64 tons 13 cwt	65 tons 4 cwt	66 tons 6 cwt
Axleload:	17½ tons	17 tons	17½ tons
Tank capacity:	1,850 gallons	1,880 gallons	1,900 gallons
Tractive effort	17,116lb	17,116lb	17,116lb

Parallel boilers were originally fitted to No.11 and 3901-3620 and a short coned boiler to 3621-3630. Long coned boilers were introduced in 1905 and replaced the earlier boilers when new boilers were required. Boiler pressure was later increased to 195lbs psi, raising the tractive effort to 18,542lb and there was experimentation with increased cylinder diameter as five engines received 17½inch cylinders between 1906 and 1909, 3617 reduced the diameter to 16 inches in 1908 and 3610 increased further to 18 inches in 1913. Extended bunkers were fixed to the last ten and 3607, 3613, 3616, 3618 and 3620 acquired them later. No.11 was renumbered 3600 in 1912. All were superheated commencing in 1912 but the last, 3611 and 3621, were not superheated until 1927. Initially, this did not require smokebox extensions, though some acquired them from 1912 onwards, though most not until the mid-1920s. The original cast-iron chimneys were replaced with copper-capped chimneys after 1907 and after 1923 taller cast-iron chimneys reappeared. Twenty of the class were fitted with ATC somewhat belatedly between 1928 and 1931 only three or four years before their withdrawal.

Initially, around fifteen of the engines worked suburban services in the Birmingham area and most of the others in the London District although a few were at Cardiff and Neath. The County 4-4-2Ts covered the faster London suburban trains with the 36XX and Metro Tanks hauling the slower stopping trains. They were displaced in the Birmingham area by the 31XX Churchward 2-6-2Ts and the last ones in the London area by the Collett 61XX 2-6-2Ts in 1931. Six were working for a short time in Devon in the first decade of the new century. In the 1920s, a significant number were at Chester for the Birkenhead services. They were all withdrawn between 1930 and 1934, the last to go being 3604, 3610, 3618 and 3628 in November 1934. Most achieved around half a million miles in their thirty year career, the highest, 3617, reaching just under 800,000 miles.

The prototype built in 1900 and thought to be running in from its delivery to traffic at Swindon at Gloucester, 1900. Note the 'guivering' of the tank side and burnishing of the buffers. (W. Vaughan-Jenkins/F.K. Davies/John Hodge Collections)

The prototype No.11 now equipped with copper capped chimney, in the Birmingham area, c1908. (Bob Miller/MLS Collections)

3604 built in 1902 with parallel boiler and cast iron chimney and 'guivered' by its crew, c1903. (H. Gordon Tidey/Bob Miller/MLS Collections)

***Above left**: **The first** of the production run, 3601, built in February 1902, but seen here after being fitted with a long coned boiler, copper-capped chimney, superheating and extended smokebox, c 1920. (Bob Miller/MLS Collections)*

***Above right**: **The last** engine of the first production run, 3620, built in 1902. It is fitted with parallel boiler, cast iron chimney and first style bunker, c1910. (LPC/F.K. Davies/John Hodge Collections)*

***Below**: **3616 in** final form with long cone boiler, tall cast-iron chimney, superheated and with extended smokebox, at Stafford Road, c1930. (Bob Miller/MLS Collections)*

Standard Gauge Locomotives built by the GWR before 1923 • 73

Above left: The prototype 3600 (the former No.11) built in 1900 but now superheated with long cone boiler, and extended smokebox, at Old Oak Common, 16 May 1925. (Bob Miller/MLS Collections)

Above right: 3622, one of the Churchward 1903 locomotives with the larger bunker, long cone boiler and extended smokebox but still with copper-capped chimney, at Old Oak Common, c1925. (W.L. Good/Bob Miller/MLS Collections)

Below: 3601, immediately after its parallel boiler was replaced by a long-coned boiler, but before being superheated or acquiring an extended smokebox, with an express, possibly from Oxford, with a rake of maroon clerestory stock on the Up relief line between Reading and London, 1909. (F. Moore/Bob Miller/MLS Collections)

Above left: **3615 with** parallel boiler in initial lined green livery at Swindon, 15 October 1904. (W. Beckerlegge/F.K. Davies/John Hodge Collections)

Above right: **3612 departing** from St. Fagans with a Cardiff – Swansea stopping train, 1923. (H.T. Hobbs/F.K. Davies/John Hodge Collections)

Below: **3628 of** the second series with coned boiler, extended smokebox and high bunker with a set of Dean 6-wheel coaches pulls away from Stourbridge Junction with a stopping train from Wolverhampton, c1923. (LGRP/F.K. Davies/John Hodge Collections)

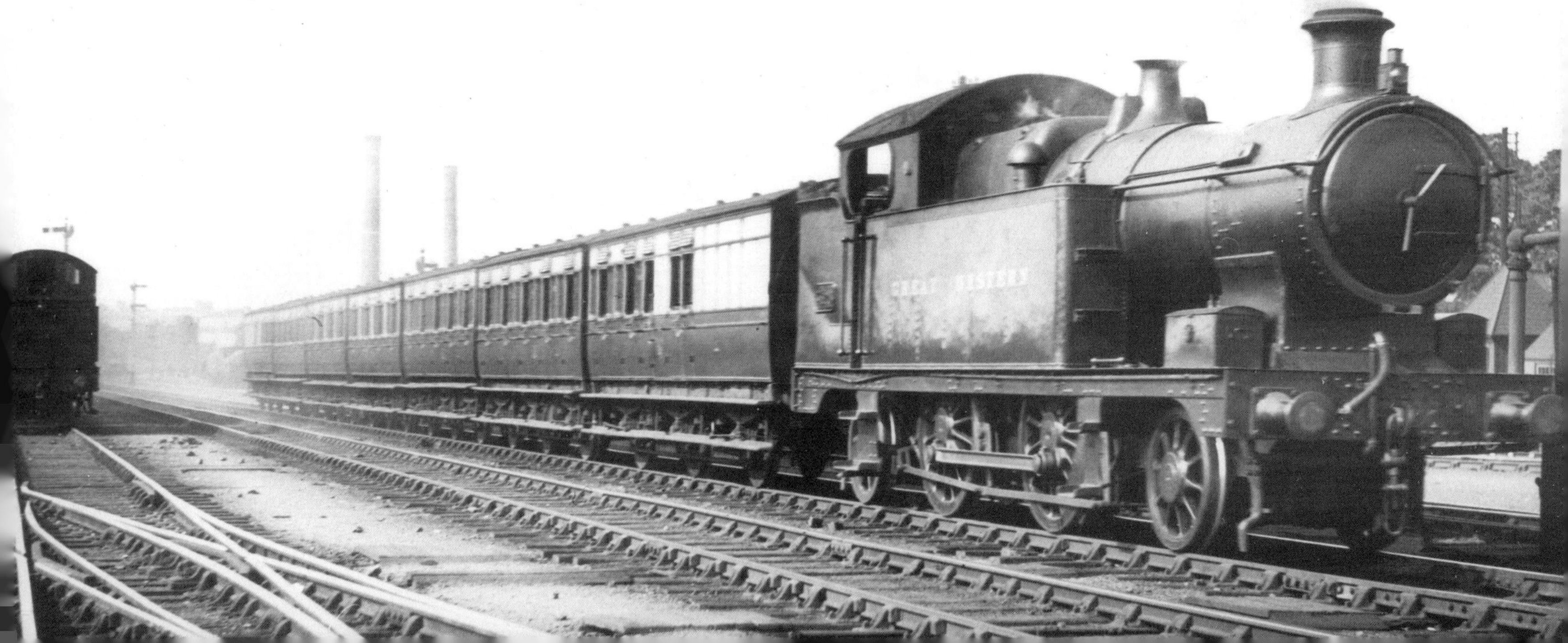

Above left: **3625 with** long coned boiler and extended smokebox with a suburban train in the West Midlands, c1925. (LGRP/MLS Collection)

Above right: **One of** the locomotives based at Chester in the 1920s, 3629, arriving at Capenhurst with a Birkenhead-Chester stopping train. 3629 was withdrawn in 1931. (F. Moore/Bob Miller/MLS Collections)

Below: **3627 departs** from Birkenhead Woodside in a flurry of leaking steam with a train for Chester, 1930. (J.A. Coltas/MLS Collection)

3616 still with its 1902 style of bunker, at Leamington Spa, 1931.
(G. Coltas/MLS Collection)

3629 at Hooton with a Birkenhead-Chester stopping train, c1930.
(H. Gordon Tidey/MLS Collection)

2221-2250 'County Tank' 4-4-2T, 1905

An account of the design, construction and operation of the County Tanks and the smaller wheeled 4600 class was included in my Pen & Sword book on the 'GW County' classes published in 2018 and is repeated here for completeness. For those wishing to see more illustrations, please see that book.

Design & Construction

The first Churchward 4-4-2 tank engine was built at Swindon in 1905. Despite the success of the 2-6-2T, No.99, and its production successors, the 3100 class, the new Locomotive Superintendent included a 6ft 8½in 4-4-2T in his range of standard locomotives for the GW. Initially, the 3100 class were seen as freight and banking locomotives, and the new engine numbered 2221 was conceived as its passenger equivalent, the tank engine version of the outside cylinder 4-4-0 'County' class, the first ten of which had been constructed the previous year. Inevitably, these new engines became known as the 'County Tanks'. At the time, the suburban trains from Paddington to Reading, Newbury and Oxford were the preserve of the 2-4-0 'Metro Tanks' and the ungainly 2-4-2T '3600 class'. With their large diameter coupled wheels – very unusual for a tank engine – they were clearly intended for the longer-distance semi-fast suburban trains. This is clear from their provision with water pick-up apparatus (two-way as they were expected to work bunker-first) although these were removed in the 1920s. As the 31XX engines with similar standard parts had already been comprehensively and successfully tested, there was no prolonged testing of 2221 before the production batch was constructed.

The main dimensions of the 'County Tank' were similar to the 'County' 4-4-0. As well as the 6ft 8½in coupled wheels, the bogie wheels were 3ft 2in. The trailing pony truck wheel diameter was 3ft 8in. The two outside cylinders were

One of the first batch of ten built in 1905-6, 2227. (F. Moore/MLS Collection)

18in x 30in stroke as for the 4-4-0. The tank engine was fitted with the smaller standard No.2 boiler, long-coned but short smokebox, pressed at 195lbs psi, giving a tractive effort (at 85 per cent) of 20,010lb. The grate area was 20.35sqft and total heating surface was 1,517.89sqft. Bunker capacity was three tons of coal and the tanks could hold 2,000 gallons of water. The prototype had cab flush with tanks and bunker (later engines had slightly wider tanks and bunker). The engine's axle-load was 19 tons and total weight, 75 tons.

One of the initial batch, 2230, was equipped with the heavier No.4 boiler similar to the 4-4-0s, but this increased the weight by around 3½ tons and would have increased the axle-load, and the experiment was short-lived, it being exchanged for a No.2 boiler in January 1907 having run less than 7,000 miles with the larger boiler. There was a proposal to build a 4-4-4T with a No.4 boiler and this may have been a trial, but the latter design was never progressed – however, a drawing exists. A few dimensions were indicated – 18in x 30in cylinders, coupled, bogie and trailing wheels of similar diameter to the 'County' tanks, tank capacity 2,000 gallons, weight a massive 82 tons 5 cwt and tractive effort, 20,530lbs.

The initial production batch (2222-2230) had number plates on their bunkers, and the words 'Great Western' on the tank side without the GW crest. (2221 had its numberplate on the tank sides.) They had cast-iron chimneys and 2221 had steam sanding gear although this was removed in 1906. A second batch (2231-2240) was built in 1908-9, which differed by being fitted with copper-cap chimneys and larger vacuum cylinders. Some of the earlier engines had received copper-cap chimneys by this time and all were now receiving the livery including the GW crest. 2225 suddenly appeared in a chocolate lake livery like some of the GW coaches at this time and was transferred to the Bristol/Swindon area for further trials.

Superheating for these locomotives, as for many of the GW's locomotive fleet, commenced in 1910 and was completed by 1914. A final batch, 2241-2250, was built in 1912, already superheated. These had curved front footplating, top feed and extended smokeboxes. The superheated engines had a reduced total heating surface of 1,316.14sqft, although other detailed differences between the last ten and the first twenty locomotives remained. Boiler pressure was standardised at 200lbs psi from 1919, increasing the tractive effort to 20,530lbs and bunker tops were extended backwards between 1922 and 1925, some getting recessed fenders at the rear to hold the upper lamp iron. Because these engines ran over the London area main lines, some received ATC apparatus early in 1908 and most others around 1915-16. Most engines in the last years before withdrawal got cast-iron chimneys again.

The construction of the 61XX 2-6-2Ts with 225lbs psi boilers in 1931 by Charles Collett for the London suburban services marked the end of the 2221 class and sixteen of the class had been withdrawn by October 1932. The last in service were 2235, 2242 and 2246 at Reading, condemned in January, September and November 1935 respectively. 2243, withdrawn in December 1934, was retained

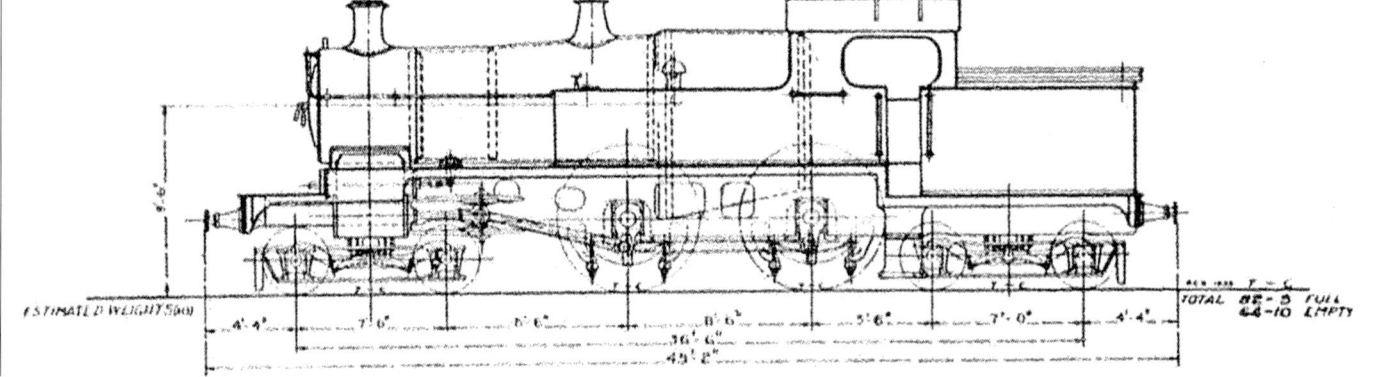

This proposal was for an extended version of the 'County' tank with larger bunker, but although a drawing exists, only a few outline dimension details have been found. Swindon drawing.

at Old Oak Common to perform carriage heating duties and was not sent to Swindon for scrapping until August 1939. Locomotive ages ranged from twenty to thirty years and total mileage between 583,000 and 876,000, and they were withdrawn because of the appearance of more suitable locomotives for the work, rather than life-expiry. The size of wheels made them capable of fast running, but their lack of adhesion and large wheels made them less suitable for frequently stopping services and the heavier trains as commuting into London increased.

Operation

Apart from 2221, which spent its first six months being tested in the Bristol-Swindon area, the initial batch all went to the London motive power division to cover the faster suburban services to Reading, Newbury and Oxford, and the High Wycombe-Princes Risborough/Oxford/Aylesbury route. In 1906, 2222-2230 were all working from the brand new Old Oak Common depot, although 2225 moved to Reading for three months in 1907, whilst two of the original batch were transferred to Slough in 1909, and a further one in 1910. The 1908-9 batch were spread between Old Oak Common, Slough and Reading, with the majority at the London depot. Nine of the new 1912 built batch went to Old Oak Common that year, with some of the older engines moving to Slough and Reading. By 1912, Old Oak had fourteen on its books, increasing to seventeen in 1913 and maintained a fleet between twelve and fifteen engines until 1920, when the average allocation there fell to ten or twelve, with four going to the Wolverhampton division.

In 1902, slip working was introduced on some London suburban services with coaches for Windsor being slipped at Slough from services bound for Henley or Oxford. The joint line with the Great Central to Denham, Gerrards Cross and High Wycombe was open for suburban services from 2 April 1906, and the new 'County Tanks' allocated to Old Oak Common would have found immediate employment on these as well as the faster services to Henley and Reading. There were five new services each way from Paddington to Aylesbury and five to Oxford via Princes Risborough and Thame. After the Birmingham route via Bicester was opened in 1910, a morning and evening service to Banbury was implemented with either 'County' or 'County Tank' haulage, allowed just 30 or 31 minutes for the 26.4 miles from Paddington, mainly against the grade and faster than the 9.10am express two-hour Birmingham train, which was allowed 32 minutes. The morning commuter services from Gerrards Cross were allowed 28 minutes for the 17.5 miles to Paddington, 23 minutes from Denham (15 miles), 18 from Ruislip (11.8 miles), and 13 minutes from Greenford (7.8 miles) involving some sharp start-to-stop times on these falling gradients before slowing for Old Oak Common and the approach to Paddington station. The evening services – against the gradient – were also tightly timed, with a 12 minute allowance to Greenford, 20 to Ruislip, 24 to Denham, 30 to Gerrards Cross and 38 minutes to Beaconsfield (21.5 miles). By 1913, when these services would have been monopolised by the 'County Tanks', there were five trains each way between Paddington and Henley, slipping coaches at Slough for the Windsor branch plus two evening services for Oxford which slipped coaches at Slough for Taplow, and Maidenhead for High Wycombe. In the book *Great Western London Suburban Services* by Thomas B. Peacock (reprinted by the Oakwood Press in 1970) the author states that 'the "County Tanks" were responsible for outstanding performance on fast heavy trains between Paddington and Windsor, Henley and Reading.'

Over the years, a few examples were tried in different locations, although the London area always remained their main sphere of operation until their replacement by the 61XX class. Other locations at which they worked, mainly for short periods, were:

Trowbridge/Westbury:	2221 (1906-18), 2228, 2237 for short periods (working locals to Westbury/Bristol/Hungerford/Reading)
Swindon:	2221 (1919), 2222, 2229, 2230 (short periods in 1907-8 and 1915)
Newton Abbot:	2226, 2236 (very short periods in 1907 and 1911)
Newport:	2230 (1907-08), 2250 (two months in 1918)
Landore:	2224 (two months in 1911)
Wolverhampton:	2223/25/26/30/42/43/46 (for 18 months in 1920-21, replaced by 31XX)
Chester:	2246 (two months in 1920)
Taunton:	2224 (four months in 1921)

Most of these trials appear to have been unsuccessful, the work being more suitable for the increasing number of 5ft 8in wheeled 2-6-2Ts. To give a snapshot, the allocation at the amalgamation of the 'Big Four' at the end of 1922 was as follows:

2221	Slough	2236	Reading
2222	Slough	2237	Slough
2223	Slough	2238	Old Oak Common
2224	Slough	2239	Aylesbury
2225	Aylesbury	2240	Reading
2226	Slough	2241	Old Oak Common
2227	Old Oak Common	2242	Slough
2228	Slough	2243	Slough
2229	Slough	2244	Reading
2230	Aylesbury	2245	Old Oak Common
2231	Reading	2246	Slough
2232	Slough	2247	Old Oak Common
2233	Old Oak Common	2248	Slough
2234	Slough	2249	Slough
2235	Aylesbury	2250	Slough

Many of the faster suburban services, including the slip-coach working, were suspended due to the First World War between 1917 and 1919 and the few remaining slip coaches still running on suburban trains in 1939 were suspended at the start of the Second World War and never restored. The 'County Tanks' resumed their fast train working, especially after 1922, with the 4.40pm Paddington-Banbury accelerated by a couple of minutes to Denham, Gerrards Cross and Beaconsfield, although in the late 1920s double-framed 4-4-0s and the 'County' 38XX would take over the Oxford via Princes Risborough and the Banbury train. Half the class was based at Slough by this time and none of the 'County Tanks'

2235 at Paddington platform 9 with an empty stock train for Old Oak Common, 1932.
(G A. Coltas/MLS Collection)

2240, the last of the second batch of 'County Tanks', built in 1908-9, at Hayes with a Paddington-Reading-Oxford semi-fast, c1925. (Photomatic/MLS Collection)

were then outside the London division. At the end Didcot received three in 1933/4 and Oxford just one (2225) in 1934, but by this time most of the Old Oak and Slough engines had been replaced by the Collett 61XX 'Prairie Tanks' and had been withdrawn. The last three (2235, 2242 and 2246) were withdrawn from Reading depot during 1935.

Despite searching for logs of performance of the 'County Tanks' I can trace no records either in the 'Locomotive Practice and Performance' articles of the *Railway Magazine* or in the archives of the Railway Performance Society. They appear to have been used mainly on the semi-fast services in the London area, with the all-stations stopping trains remaining by and large the duty of 'Metro' 2-4-0Ts, the 36XX 2-4-2Ts, and the 39XX 2-6-2Ts (the rebuilt 'Dean Goods', some of which transferred to the London area in the late 1920s) until the 1931 built 61XX replaced all of them. With their large driving wheels and moderate adhesion, they would have struggled to accelerate heavy stopping trains, as I discovered in 1958 when I travelled on an eight-coach suburban train (the 6.20pm Paddington-Reading) behind the 6ft 8in 4-4-0, *City of Truro*. However, once on the move they would – like their tender 4-4-0 counterparts – have displayed a good turn of speed and I'm sure speeds in the 65-75mph range between Paddington and Reading, between Reading and Didcot, or especially between Beaconsfield and Ruislip, would have been commonplace, even if the riding was a bit rough, as apparently they shared that reputation with the 38XX 'Counties'.

4600, the small-wheeled (5ft 8in) 4-4-2T, in Works grey, 1913. This was the only example built. (MLS/Moore's postcard)

4600, 4-4-2T, 1913

Design, Construction & Operation

A smaller wheeled 4-4-2T was constructed at Swindon in November 1913. It was a version of the 2-6-2 smaller passenger tank engine (the 45XX class) and was numbered 4600. There were still a few 'metro 2-4-0' tanks being used on branch line work at this time and their replacement by the 4-4-2T was considered, though the success of the 45XX, the start of the First World War and increasing road competition for branch traffic meant that no further locomotives of this type were required. There was also a need for engines to work growing suburban traffic round Bristol and Birmingham – as distances between stations were shorter than London, it is assumed that smaller wheels were considered appropriate. However, the success of the 31XX 2-6-2Ts covered this area of developing work.

Its dimensions were 5ft 8in coupled wheels (larger than the 45XX and intended therefore to increase maximum speeds), with bogie wheels of 3ft 2in diameter, and trailing wheels of similar size. Its boiler was pressed at 200lb psi, with grate area of just 16.6sqft and total heating surface of 1,271.86sqft. Two cylinders of 17in x 24in stroke were provided and the weight was 60 tons 12 cwt, significantly lighter than the 'County Tank', with just 16 ton axle-load. Tractive effort (at 85 per cent) was 17,340lb. The tank capacity was only 1,100 gallons

4600 at its first allocated shed, Tyseley, shortly after delivery, c1914. (G W Trust)

indicating that short distance suburban or branch line work was anticipated.

It had curved footplating at front and rear, front end strengthening struts from the boiler saddle to buffer beam, copper-cap chimney and numberplates on the bunker side. The engine was superheated in 1918, reducing the total heating surface to 1,215.52sqft and reducing the total weight by 5 cwt.

It was initially allocated to Tyseley for suburban work in the Birmingham area. However, after the First World War and with the 31XX and 39XX 2-6-2Ts and 36XX 2-4-2Ts dominating the suburban trains both north and south of Birmingham, 4600 took up rural duties based at Neyland for the Pembroke Dock branch, where it remained until withdrawal in July 1925.

1-99 Steam Railmotors, 0-4-0T, 1903

The first Great Western rail motors were built in 1903 and were multiplied rapidly, so that ninety-nine were in service by 1908. The majority of the locomotive portions and coach bodies were built at Swindon but fourteen were built by Kerr, Stuart & Co., and twenty at Stafford Road Works, Wolverhampton. Although they were allotted engine numbers ranging from 0801 to 0942 the units were known by their coach numbers 1-99. The engine portions were not permanently linked to the same coach as changes could take place when the engine needed more time in overhaul than the trailing coach element and swaps took place. There were 112 engine portions built for the 99 trailers, thus there were 13 spare loco units. The power units were steam engines on an 0-4-0T layout and had the broad common basic dimensions:

Coupled wheels:	4ft 0in (0801-5, 3ft 6in)
Cylinders:	12in x 16in
Boiler pressure:	160lb psi
Heating surface:	600-625sqft (0803-5 only 436.72sqft)
Grate area:	11.54sqft (0803-5 8.4sqft)
Weight (engine):	22-26 tons (Locos for units 84-99 – 29 tons)
Weight (total):	36-43 tons (Total for units 84-99 – 45 ½ tons)
Tank capacity:	450 gallons
Seating capacity:	49-63 (majority 50-52)

Engines 0864 and 0865 married with railmotor trailing cars 15 and 16 were the first two designed and built by Kerr, Stuart & Co., and were considerably less powerful with different dimensions, being:

Coupled wheels:	3ft 5in
Cylinders:	9in x 15in
Boiler pressure:	170lb psi
Heating surface:	350.5sqft
Grate area:	7.25sqft
Weight (Total):	32 tons
Tank capacity:	338 gallons
Seating capacity:	48

The rest of the Kerr, Stuart built railmotors worked broadly to the GW design. In their early days they were painted crimson lake including wheels, although later when the coach was chocolate and cream the wheels were black. They were most suited to rural branch lines as on more populated areas they were victims of their own success, and the passenger numbers outgrew their capacity. By 1905, the GW was matching auto trailers with engines of the '517' 0-4-2T class and after 1907 with some 'Metro Tanks' also as these could be strengthened with extra trailers to carry the increasing number of passengers. This meant the withdrawal of the first railmotors by 1914 and then a gradual withdrawal until the last, Nos. 30, 37, 55, 65, 70, 71, 88, 91, 92 and 97 in 1935. No.93 with Kerr, Stuart loco part 0873 was withdrawn in December 1934 and subsequently preserved and rebuilt by the GW Society at Didcot.

The following services were operated from 1905 onwards at one time or another by railmotors fully or in part:

London Division
    Southall-Brentford
    West Drayton-Uxbridge & Staines
    Gerrards Cross-Uxbridge
    Westbourne Park-Ruislip & Southall
    Slough-Windsor
    Park Royal-Willesden Junction
    Reading-Henley & Maidenhead & Didcot
    Oxford-Princes Risborough

Bristol Division
    Bristol-Clifton Down & Avonmouth & Portishead
    Swindon-Chippenham-Bristol
    Westbury-Castle Cary-Taunton

84 • FOUR-COUPLED TANK LOCOMOTIVE CLASSES BUILT BY THE GREAT WESTERN RAILWAY

Westbury-Chippenham & Patney & Warminster
Trowbridge-Devizes
Limpley Stoke-Hallatrow

Yatton-Clevedon & Highbridge & Wells
Weymouth-Dorchester & Abbotsbury

Newton Abbot Division
    Exeter-Dulverton & Dawlish Warren & Heathfield
    Plymouth-Saltash & Yealmpton & Tavistock
    Lostwithiel-Fowey
    Taunton-Milverton
    Truro-Newquay

Wolverhampton Division
    Banbury-Princes Risborough & Chipping Norton
    Rock Ferry-Ledsham
    Wrexham-Rhos & Coedpoeth & Moss & Llangollen
    Much Wenlock-Craven Arms
    Stourbridge Junction-Stourbridge Town & Wolverhampton
    Dudley-Old Hill
    Langley Green-Oldbury
    Old Hill-Halesowen
    Tyseley-North Warwick line

Worcester Division
    Chalford-Stonehouse (the first route in 1903)
    Cheltenham-Honeybourne
    Newnham-Forest of Dean
    Kidderminster-Bewdley
    Honeybourne-Stratford-on-Avon & Evesham

Newport Division
    Pontypool Road-Monmouth & Oakdale
    Aberdare (LL)-Dare Valley
    Merthyr-Newport

Neath Division
    Neath-Court Sart
    Glyn Neath-Swansea East Dock
    Pembroke Dock-Tenby
    Garnant-Gwaun-cae-Gurwen
    Fishguard Harbour-Clarbeston Road

The first Great Western Steam Railcar No.1 at Chalford, 1903. (M.G.D. Farr Collection)

Great Western Steam Railcar No.1 leaving Brimscombe Bridge on trial on the Gloucester-Chalford service, 1903. (M.G.D. Farr Collection)

A GW steam rail motor No.59 and trailer, c 1912. (GW Trust)

One of the last GW steam rail motors to remain in operation, No.71, in the early 1930s. (GW Trust)

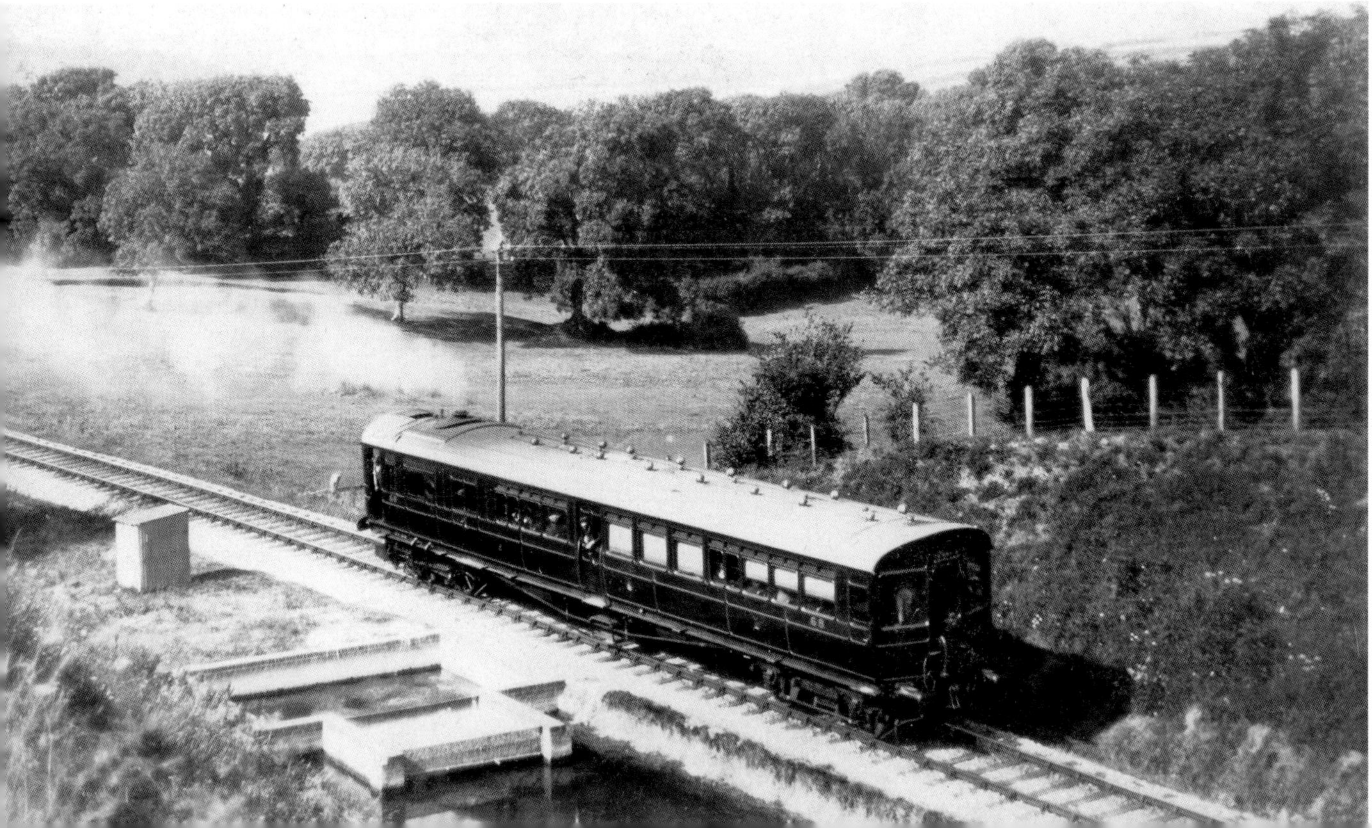

GW steam rail motor No.68 in operation on a GW branch line, c1920. It was withdrawn in 1922. (GW Trust)

GW steam rail motor No.70 at Southall in the 1930s. (John Hodge Collection)

GW steam rail motor No.48 at Lamphey with a Whitland – Pembroke Dock branch train, hauling a 4-wheel trailer and box wagon, c1930. (John Hodge Collection)

A pair of GW steam rail motors with a couple of trailers inside at Upper Sudeley Halt in the Forest of Dean in the late 1920s. (John Hodge Collection)

Experimental locomotives

1, Dean 4-4-0T, 1880, reb.2-4-0T, 1882

William Dean's first design, delivered in 1880, appropriately numbered 1, was a strange 4-4-0 side tank with double frames and a novel framing for the four-wheeled front truck whose frames were independent of the frames of the rest of the engine. The boiler was standard Swindon type and its dimensions were:

Coupled wheels:	5ft 6in
Leading truck wheels:	3ft 9in
Cylinders:	17in x 26in
Boiler pressure:	140lb psi
Heating surface:	1,188.6sqft
Grate area:	19sqft
Weight:	43 tons 8 cwt
Axleload:	15 tons 16 cwt
Tank capacity:	935 gallons
Tractive effort:	13,548lb

The engine was clearly a problem as it was rebuilt after less than two years, after only running 2,870 miles, as a more conventional 2-4-0 side tank. Because of its short life in original form, no photos of it have been found, but a drawing made by Mr E.W. Twining in *The Locomotive* magazine of January 1940 is shown below.

Above: Dean's experimental 4-4-0T rebuilt in 1882 as a 2-4-0T. (F. Moore/Bob Miller/MLS Collections)

Left: Dean's experimental 4-4-0T rebuilt in 1882 as a 2-4-0T and further rebuilt with Belpaire boiler in 1914, seen here c1920. (Bob Miller/MLS Collections)

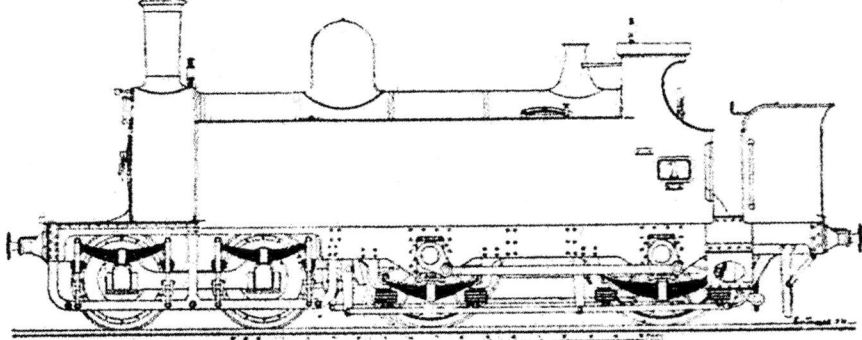

The rebuilt 2-4-0T emerged in May 1892 with a new front end with normal outside bearings and shortened tanks. Its revised dimensions were:

Leading wheels:	4ft 0in
Boiler pressure:	165lb psi
Heating surface:	1,307.83sqft
Weight:	47 tons 8 cwt
Axleload:	17 tons

Dean's No.1 at Chester in March 1920. (J.N. Maskelyn/F.K. Davies/John Hodge Collections)

Tank capacity: 882 gallons
Tractive effort: 14,920lb

In this new form it was successful and was reboilered in 1914 and given an extended smokebox and top feed. In its original form it worked in the Bristol area and remained there after rebuilding, working to Westbury, Salisbury and Taunton. Then in the new century, after a short stay in Plymouth, it moved north to Chester and lasted until 1924, running just over 530,000 miles in total.

13, Dean 2-4-2T, 1886, reb.4-4-0T, 1897

Dean designed and built another experimental locomotive in 1886, which was numbered 13. It was the first 2-4-2T built by the GWR and had tanks at the back and well tank underneath. The coupled wheels had inside and the leading and trailing wheels outside bearings. Its other dimensions were non-standard to other GWR practices:

Coupled wheels: 4ft 0in
Leading & trailing wheels: 3ft 6in
Cylinders: 16in x 21in
Boiler pressure: 140lb psi
Heating surface: 864.06sqft

Dean's experimental 2-4-2T, No.13, as built in 1886. (F. Moore/MLS Collection)

Grate area: 11.37sqft
Weight: 33 tons 16 cwt
Axleload: 9 tons
Tank capacity: 600 gallons
Tractive effort: 13,328lb

It appears to have been intended for local branch work and worked on both the Abingdon and St Ives branches before being rebuilt in 1897 as a 4-4-0 saddle tank, with the following revised dimensions:

Coupled wheels: 4ft 1½in
Bogie wheels: 2ft 8in
Weight: 35 tons 19 cwt
Axleload: 11 tons 9 cwt
Tank capacity: 628 gallons

In 1901 it received a new boiler and extended smokebox. After initial use on the Swindon-Highworth branch it spent the rest of its career on the Liskeard-Looe

No.13, as rebuilt as a 4-4-0T, seen at Swindon during its time as Works shunter, c1925. Ignore the crane – it was not one of the crane engines!
(Bob Miller/MLS Collections)

Dean's experimental 2-4-2T, No.13, as rebuilt in 1897 and working from around 1900 to 1922 on the Liskeard-Looe branch, seen here at Looe carriage sidings, c1910.
(F. Moore/MLS Collection)

branch until 1922 and then a further four years until its withdrawal in 1926 as a Swindon Works shunter. Its total mileage in traffic was 681,230.

34, 35, Dean 0-4-2ST, 1890, reb 0-4-4T, 1895

Dean designed and built two 2-4-0 saddle tanks in 1890 which diverged from the Metro Tanks and the 3521 2-4-0Ts in many ways and whose purpose was unclear. They were numbered 34 and 35 and had inside plate frames and bearings and were smaller than the 3521 class, and presumably suffered from the same unsteadiness as the 2-4-0T version of the 3521 class because they were similarly rebuilt as 0-4-4 tanks in 1895. Their original dimensions were:

Coupled wheels:	4ft 0in
Trailing wheels:	3ft 6in
Cylinders:	16in x 24in
Boiler pressure:	150lb psi
Heating surface:	1,235.2sqft
Grate area:	14.8sqft
Tank capacity:	950 gallons
Tractive effort:	16,320lb

After rebuilding, their altered dimensions were:

Coupled wheels:	4ft 1½in
Trailing bogie:	2ft 8in
Weight:	40 tons 6 cwt
Axleload:	12¼ tons
Tank capacity:	850 gallons
Tractive effort:	15,825lb

Dean's 34 rebuilt in 1895 as a 0-4-4T, seen in the West Country, c1900. (F. Moore/MLS Collection)

Dean's 34 rebuilt in 1895 as an 0-4-4T, sold in 1908 to the Woolmer Instructional Military Railway and named *Longmoor*, the location of the railway, c1910. (MLS Collection)

Like the larger 0-4-4Ts, they spent their time in the West Country and as 0-4-4Ts worked the Helston and St Ives branches. 34 was later fitted with an enclosed cab and was sold in 1908 to dealers in secondhand locomotives, finishing at Longmoor Military Railway and bearing the name *Longmoor*. It was scrapped at Swindon in 1921 having been sent for repair which was found not worth doing. 35 was withdrawn in 1906.

1490, Dean 4-4-0PT, 1898

In October 1898 the GWR constructed a singular 4-4-0 pannier tank intended initially to replace the 'Metro' tanks on the London suburban services but, weighing over fifty tons with an axleload of 17 tons 12 cwt, it was too heavy for that use and additionally it was found to be too unstable at passenger train speeds. Although designed and built in the Dean era, it had an early Belpaire firebox indicating Churchward's growing influence and requiring redesign of the tanks away from the saddle variety which could not easily accommodate the raised firebox. It had two 15½ in x 26in cylinders, coupled wheels of 4ft 7½ in, boiler pressure of 165lb psi, with a total heating surface of 1,484.34sqft., and a grate area of 20.41sqft. The tank water capacity was 1,075 gallons and it had a tractive effort of 15,785lb.

Unsuccessful for the duties for which it was designed, it was not replicated, but relegated to freight and shunting work in the Bristol/Gloucester area at first, then later taking up the position of regular pilot/shunting engine at Bath. It was finally at Swindon before being sold in 1907 to the Ebbw Vale Steel, Iron & Coal Company, where it spent a year at Pontypool and Abercarn (named *Dickinson*) before being resold to the Brecon & Merthyr Railway in 1908 and later resold again to Cramlington Colliery Company in Northumberland. The latter company made a number of modifications intended to make it more suitable for heavy colliery traffic, but it was scrapped in 1929.

Whilst clearly not very successful – indicated by the number of owners who got rid of it – it has the distinction of being the first GWR engine to adopt the pannier tank form, in this case starting only at the back of the smokebox, in similar fashion to Hawksworth's 94XX and 15XX in the 1940s.

Portrait of Dean's 4-4-0PT No.1490, as built, c1899. (F. Moore/MLS Collection)

Brecon & Merthyr 4-4-0PT, formerly GW 1490, acquired in 1908, at New Tredegar (White Rose), c1910. (R.C. Riley/F.K. Davies/John Hodge Collections)

GW 4-4-0PT 1490 at Slough on Paddington suburban duties for which it was designed and found wanting, at Slough, c1900. (MLS Collection)

101, oil burning 0-4-0T, 1902

The final experimental four-coupled tank described was an 0-4-0 side tank built as an oil burner using the G.E. Holden system. Although conceived in Dean's time, 1902, it was not built until 1903 after Churchward had taken command and it is probable that it was his initiative as Dean was already in poor mental health from the turn of the century. It was numbered 101 and its dimensions were:

Coupled wheels: 3ft 8in
Cylinders: 13in x 22in

Joy's outside valve gear
Boiler pressure: 180lb psi
Heating surface: 907.83sqft
Grate area: 10.9sqft
Fuel capacity: 200 gallons

The firebox chamber length was reduced from 4ft 10in to 3ft 1½ inches in the initial stages and it received a Lentz boiler designed in Prussia, which Churchward may have discovered on one of his American trips as it was found in the Vanderbilt designs of 1899. It was tested for a year but was taken out of service in 1904 and rebuilt to burn coal in 1905. Its final revised dimensions were:

Boiler pressure: 160lb psi
Heating surface: 825.18sqft
Grate area: 7.78sqft
Weight: 28 tons 13 cwt
Axleload: 15½ tons
Tank capacity: 500 gallons
Tractive effort: 11,492lb

It was confined to Swindon Works as a shunter there and withdrawn in 1911.

Former GW 4-4-0PT 1490 sold to the Cramlington Colliery Company in the 1920s, rebuilt with a substantial and unique enclosed cab. (R.H. Inness/F.K. Davies/John Hodge Collections)

A portrait of 101 as delivered in 1903 as an oil-burning experimental 0-4-0T. (MLS Collection)

Chapter 4
LOCOMOTIVES BUILT BY THE GWR AFTER 1923

1101-1106, 0-4-0T, 1926

The Great Western Railway inherited a large number of locomotives in a poor state of repair from the South Wales absorbed companies in 1922 and after undertaking an initial flurry of necessary repairs, Collett decided that it was prudent to replace some of the oldest with standard designs. Swindon was fully occupied with building the Castles and then the Kings in the mid-1920s, so the construction of many of the 0-6-2T 56XX locomotives designed to replace the myriad classes of Victorian tank engines was outsourced to private companies. The docks in South Wales were expanding to cope with the high tonnages of coal for export and some of the oldest shunting engines on Swansea Docks had been sold before the First World War and

1104 in original condition with square topped cab on Swindon shed with another of the class after delivery by the Avonside Engine Company of Bristol, 1926. (Bob Miller/MLS Collections)

Locomotives built by the GWR after 1923 • 95

1105 still in GWR livery with the rounded cab, GW safety valve cover and bell in front of the cab used to give warning of its approach to dockside traffic, at Danygraig shed, 31 July 1950. *(MLS Collection)*

1105 now in BR livery of black with red backed numberplate at Danygraig shed, 1953. *(N. Harrop/MLS Collection)*

An 1101 class Avonside Engine Co. 0-4-0T of 1926 glimpsed on the author's walk from Swansea East Dock to Danygraig as the 6am shift started on a grey August morning, 1957. (David Maidment)

some of the Peckett engines of the Swansea Harbour Trust and the Powlesland and Mason Company were near the end of their economic lives, so six modern 0-4-0 side tanks to the Avonside Engine Company's standard design were ordered and delivered by that company in 1926. They were numbered 1101-1106 and were known as the '1101' class. Their dimensions were:

Coupled wheels:	3ft 9½in
Cylinders:	16in x 24in
Walschaerts valve gear	
Boiler pressure:	170lb psi
Heating surface:	864sqft
Grate area:	12.57sqft
Weight:	38 tons 4 cwt
Axleload:	19 tons 8 cwt
Tank capacity:	1,000 gallons
Tractive effort:	19,510lb

They were powerful beasts for so small an engine, though with an unusually heavy axleload for a shunting engine designed to work over sidings that were not necessarily laid to the best standards. However, they were permitted over most of the lines around Swansea Docks and spent all their lives at either Swansea East Dock or Danygraig depots. The only change was the replacement of the square topped cabs with lower rounded ones, presumably to improve clearances around some of the dock lines. They continued to work on the docks until replaced by BR diesel shunting engines at the end of the 1950s and 1101 was withdrawn in November 1959 and the rest in February 1960, surprisingly outlived by three of the former SHT and P&M Peckett engines (1143, 1151 and 1152).

12-13, Sentinel Waggon Works 0-4-0T, 1926

At the same time that Collett ordered six shunting engines of the 1101 class, he decided to test two examples of the Shrewsbury Sentinel Waggon Works Ltd. They were the company's patent geared steam locomotives with vertical engines and boilers with the firebox within the boiler shell. They had poppet valve gear and a maximum speed of 18mph. One attraction was their light weight and ability

The Shrewsbury Sentinel Waggon Works, GW No.13, at Park Royal Trading Estate, 3 June 1933. (P.J.T. Reed/F.K. Davies/John Hodge Collections)

to traverse any section of track on the GWR. They were allocated the numbers 12 and 13. Their dimensions were:

	No.12	No.13 (where different)
Wheels:	2ft 6in	
Cylinders:	6¾in x 9in	
Boiler pressure:	275lb psi	
Heating surface:	71.5sqft	54.43sqft
Grate area:	5.1sqft	3.97sqft
Weight:	22 tons	20 tons
Tank capacity:	350 gallons	
Tractive effort:	7,200lb	

No.12 was intended for branch passenger work with steam heating pipes and vacuum brakes. It was actually the second of the two but the first was not taken into GW stock until trials had been completed, so receiving its number, 13, later. 12 was tested on the Fowey branch in the autumn of 1926 but found unsatisfactory and returned to the makers, and was subsequently sold to Messrs T.E. Gray of Burton Latimer and named *Isebrook*. No.13 became the GW shunting engine on the Park Royal Trading Estate or on loan to the Lyons Factory at Greenford. It was withdrawn in 1946 and sold to G. Cohen & Sons of Small Heath.

4800-4874 (renumbered 1400-1474), 0-4-2T, 1932

The 4800 class or 48XX were built between 1932 and 1936 and were a modernised version of the Wolverhampton much rebuilt '517' class of 0-4-2Ts which were being scrapped as life expired in the late 1920s and early 1930s. Costs per engine ranged from £2,247-2,285, 21 per cent of which was cost of the boiler. Some received reconditioned boilers from the class '517s'. There was criticism that they looked ancient and were little advanced on

The prototype, 4800, as built at Swindon in 1932.
(F. Moore/MLS Collection)

the 517s but in truth they did the job well enough, but their age was increasing repair costs. Collett was under pressure to reduce running costs and produced a standardised GW version with a decent cab that performed perfectly well and very economically on all the myriad branch lines throughout the GWR. Initially painted the standard GW green, by the war years they were black and were painted BR plain black apart from a couple of examples that received the mixed traffic lining early in the nationalisation years. Swindon painted a number in passenger lined green after 1957. Their dimensions were:

Coupled wheels: 5ft 2in
Trailing wheels: 3ft 8in

Cylinders:	16in x 24in
Boiler pressure:	165lb psi
Heating surface:	953sqft
Grate area:	12.8sqft
Weight:	41 tons 6 cwt
Axleload:	13 tons 18 cwt
Tank capacity:	800 gallons
Coal capacity:	2 tons 13 cwt
Tractive effort:	13,900lb

4805, as built at Swindon in 1932.
(Bob Miller/MLS Collections)

Locomotives built by the GWR after 1923 • 99

4803 at Gloucester, 13 April 1933.
(P.J.T. Reed/F.K. Davies/ John Hodge Collections)

4808 at Swindon, 9 October 1937.
(P.J.T. Reed/F.K. Davies/ John Hodge Collections)

4848, c1938.
(MLS Collection)

4805 and 4851 reside in the shed at Exeter St Davids, 22 August 1937. (F.M. Gates/GW Trust)

Locomotives built by the GWR after 1923 • 101

4870 with autocoach at Totnes, 30 August 1945. (F.M. Gates/GW Trust)

1432, as renumbered in 1946, at Oswestry shed, 1 May 1949. (MLS Collection)

1463 at Newton Abbot shed, 24 June 1958. (MLS Collection)

1424 ex-works in BR black livery, at Wolverhampton Stafford Road, 1952. (MLS Collection)

Locomotives built by the GWR after 1923 • 103

1465 in BR mixed traffic lined black livery – a livery that was rarely applied to this class – at Machynlleth shed, 20 July 1951. (MLS Collection)

1426 ex-works in the final BR lined green livery that Swindon Works applied to most ex-GWR engines after 1957, at Cardiff Canton shed, c1958. (J. Davenport/MLS Collection)

1401 during the filming of the *Titfield Thunderbolt* with the set of ancient vehicles rescued by the villagers to keep their branch line open on the Limpley Stoke-Camerton branch near Monkton Combe, 26 July 1952. (Peter Fry/ GW Trust)

The seventy-five engines of the class, numbered originally 4800-4874 and renumbered 1400-1474 in the autumn 1946 GW renumbering scheme, were all auto fitted and built with ATC equipment (at a cost of £79 per loco). When operating in auto mode, the driver had control only of regulator, brake and warning gong. The remaining controls were left to the fireman to operate, including the reverser. The firemen of auto trains were therefore frequently senior men, and it was not unknown for the regulator link to be uncoupled and the fireman to drive although this practice was not condoned by the authorities.

They were permitted a load of 260 tons on easily graded routes when not operating in auto mode and up to 168 tons when auto worked if grades no steeper than 1 in 100 (five vehicles) although only two vehicles if facing gradients as steep as 1 in 40. In the 1930s they were averaging 40,000 miles a year, although this fell after the war and by the late 1950s after the closure of some branch lines, just 20,000 miles per annum. They were neat little engines well capable of a turn of speed when the track was suitable. Their best known speed track was parallel to the Bristol-Gloucester main line on the four track section between Stonehouse and Tuffley Junction, where occasional races between the 48XX and its auto coach often got the better of a Black 5 or Jubilee on a heavier Bristol-Birmingham express. Initially the majority of the class were split fairly equally between the London, Newton Abbot, Wolverhampton and Worcester divisions, with the other Divisions having just a handful. The Cardiff Valleys Division only had one. No Division was completely left out.

At nationalisation the London Division had eleven, based at Slough (1), Aylesbury (1), Southall (1), Reading (2), Staines (2), Marlow (1), Wallingford (1), Oxford (1), and Abingdon (1). The Bristol Division had eleven also at Bath Road (1), Yatton (2), Swindon (3), Chippenham (2), and Weymouth (3). The Newton Abbot Division had twelve – at Newton Abbot itself (4), Exeter (5), Ashburton

Locomotives built by the GWR after 1923 • 105

1401 during the filming of the *Titfield Thunderbolt* in the scene in which it duels with the bus company owner's steam roller (seen on the far right), and as it lines up ready to charge Sid James' steam roller. (Steam World)

(1), and Tiverton Junction (2). The Worcester Division had the largest number – fourteen – at Worcester (2), Gloucester (4), Lydney (2), Cheltenham (1), Chalford (1), Leominster (2) and Ross-on-Wye (2). The Wolverhampton Division had eleven at Banbury (2), Stourbridge (3), Croes Newydd (5) and Penmaenpool (1). South Wales now had a fair number, with the Newport Division having three, at Ebbw Junction (1), Llantrisant (1) and Pontypool Road (1) while the Neath Division had five, at Carmarthen (1) and Fishguard (4). The Cardiff Valleys Division had three, at Cathays (2) and Abercynon (1). Finally the Oswestry Division had six – at Oswestry (4), Machynlleth (1) and Aberayron (1). Many of these allocations were at small sub-sheds to start the work in the morning from the branch end. In the mid-1950s the Bristol Division had increased their number at the expense of the Wolverhampton Division and the Cardiff Valleys engines had moved on.

1401 and 1462 both starred in the 1953 Ealing Comedy Film, *The Titfield Thunderbolt*, where 1401 was the regular branch engine run by the locals when they tried to run a closed railway (filmed on the Limpley Stoke-Camerton branch near Bath). After their engine was wrecked by sabotage by the rival bus company, the drunken volunteer driver tried to steal 1462 from a Bristol shed and a replica was perceived running along a highway before coming to grief in a collision with a tree.

I have unearthed a number of logs from the Rail Performance Society's archive in an attempt to find the highest authenticated speed by a Collett 0-4-2T and at first the best I could find there was the run below:

Stonehouse-Gloucester, 14.8.1964

9.30pm Chalford-Gloucester

1444 – Gloucester

Autocoach, 32 tons

Miles	Location	Times	Speed	
0	Stonehouse	00.00		1 L
1.8	Standish Junction	03.25	42	1½ L
	Haresfield	05.27	41	
	MP 109	07.06	57	
	MP 110	08.11	60	
	MP 111	09.10	61	
6.95	Tuffley Junction	10.13	55	
	Gloucester South Jcn	12.29	33	
	Tramway Jcn	13.30	sigs 7*	¾ L
9.15	Gloucester	15.12		1 L

Swindon's 1410 at Calne about to propel its two auto trailers on the 12.20pm Calne back to Chippenham, 10 May 1958. (James Tawse)

The Gloucester crew seem to have been motivated to get home before the pubs closed rather than racing a Midland line train. Speeds in excess of 70mph have been rumoured but I can't find the evidence for it. The outward journey of 1444 and its crew was much more sedate, holding a steady 45mph on this level section and arriving at Stonehouse a minute early in 15½ minutes for the 9.2 miles. I had a run on 17 June 1964 with 1445 and one autocoach on the 6.22pm Gloucester-Chalford which touched 50mph between Tuffley Junction and Naas Crossing and completed the run to Stonehouse comfortably within the 16 minutes allowed. My return on the 6.50pm Chalford from Stonehouse accelerated to 73mph but 2-6-2T 4100 and three coaches were substituting for the auto set. Other unsubstantiated rumours include an article in the *Railway Magazine* in the 1930s when Cecil J. Allen reported he had been overtaken by a 48XX and auto coach on the Down relief between Maidenhead and Twyford when he was in an express on the Down main, and a report of a 14XX making 70mph on the Ruabon-Llangollen line! In fact, I have just come across an article written in 1952 which includes four short logs from my friend Rodney Meadows, who was my mentor when I first joined the railway having been the officer in charge of a group of us sixth-formers at a work experience course in 1956. They are runs on the line where Cecil J. Allen says he was overtaken – four Maidenhead-Reading auto trains, the trailer in fact being propelled by the bunker-first 14XX at the rear.

Maidenhead-Twyford

7.15pm Maidenhead-Reading

14XX propelling 1 auto-trailer, 33 tons

		1444		1407		1407		1444	
Miles	Location	Times	Speeds	Times	Speeds	Times	Speeds	Times	Speeds
0	Maidenhead	00.00		00.00		00.00		00.00	
2	Waltham	-	52	-	54	-	55	-	60
3.6	Shottesbrook	-	63	-	58	-	64	-	66
5.2	Ruscombe Sdgs	-	65	-	60	-	65	05.50	67
6.8	Twyford	08.51		08.53		08.11		07.36	

Other routes for which I have logs involved frequent stops and rare opportunities to exceed 35-45mph. To give examples of the routes for which RPS members have contributed logs, I have the following:

3.50pm	Blenheim & Woodstock-Kidlington, 1448 + 110 tons, 8.2.1947. Max speed 34½mph
6.15pm	Cheltenham Spa-Honeybourne, 1413 + auto coach, 5.3.1949. Max speed 43mph
1.40pm	Radley-Abingdon, 1450 + autocoach, 17.9.1949. Max speed 32½mph
11am	Ross-on-Wye-Monmouth Troy with 1445 + auto coach, 5.8.1954. Average speed (running times only) 28mph
4pm	Cinderford-Gloucester, with 1401 + autocoach, 5.8.1954. Net average running speed – 31.1mph
10.5am	Paignton-Newton Abbot, 1452 + 65 tons, 4.9.1957. Max speed 46mph
5.48pm	Abingdon-Radley, 1435 + autocoach, 8.6.1961. Max speed 37½mph
11.40am	Chalford-Gloucester, 1472 + auto coach, 25.8.1962. Max speed 50mph

James Tawse wrote an article in *Steam World*, February 1998, about a day he spent travelling the byways of the Western Region in May 1958, which included a trip from Trowbridge to Calne and back to Chippenham with 1410 and a couple of auto trailers, 62 tons. The 11.22am Trowbridge left a couple of minutes late waiting connection off the 10.30am Bristol-Westbury, and ran the 17.1 miles to Calne in 47½ minutes including seven intermediate stops. The longest non-stop section was the 6.2 miles from Melksham-Chippenham covered in 12½ minutes including a slowing approaching Lacock Halt, a request stop, to ascertain there were no waiting passengers, so 40-45mph must have been attained. The auto train arrived in Chippenham 1½ minutes early, left a minute late and arrived at Calne ½ minute early. On the return from Calne, it left 2 minutes late and arrived at Chippenham 2 minutes early omitting the request stop at Black Dog Halt. It took 11 minutes for the 5.3 miles and again 40-45mph would have been reached.

4874 at Builth Road Low Level under the Central Wales Line from Craven Arms with a Rhayader–Three Cocks Junction auto train, 30 May 1936. (MLS Collection)

4843 on arrival at Blenheim & Woodstock from Kidlington and Oxford, 24 June 1935. (F.M. Butterfield/F.K. Davies/John Hodge Collections)

4847 rumbles along the Marlow-Bourne End branch with the 'Marlow Donkey', 17 May 1937. (J.H. Waters/F.K. Davies/John Hodge Collections)

4816 at Bearley with a Stratford-on-Avon-Leamington auto train to which freight vans have been added, 2 August 1938. (F.K. Davies/John Hodge Collections)

4849 at Dulverton with the 10.09am from Exeter St David's in the war years, 17 September 1941. (A.W. Croughton/F.K. Davies/John Hodge Collections)

Newton Abbot's long domiciled 4827 at Moretonhampstead, 15 May 1941. (A.W. Croughton/F.K. Davies/John Hodge Collections)

Locomotives built by the GWR after 1923 • 111

Gloucester's 4841 looks the worse for wear at the end of the Second World War with the Cheltenham-Honeybourne auto train, May 1946. (W. Potter/F.K. Davies/John Hodge Collections)

Newton Abbot's 1439 at Heathfield with the Moretonhampstead-Newton Abbot auto train, 3 April 1948. (C.H.S. Owen/MLS Collection)

1470 at Totnes with the auto train for Ashburton on the South Devon branch, 26 July 1951. 1470 is still displaying the pre-1948 livery three years later and it still looks in good condition.
(MLS Collection)

1470 at Ashburton with the auto train for Totnes, 26 July 1951.
(MLS Collection)

Locomotives built by the GWR after 1923 • 113

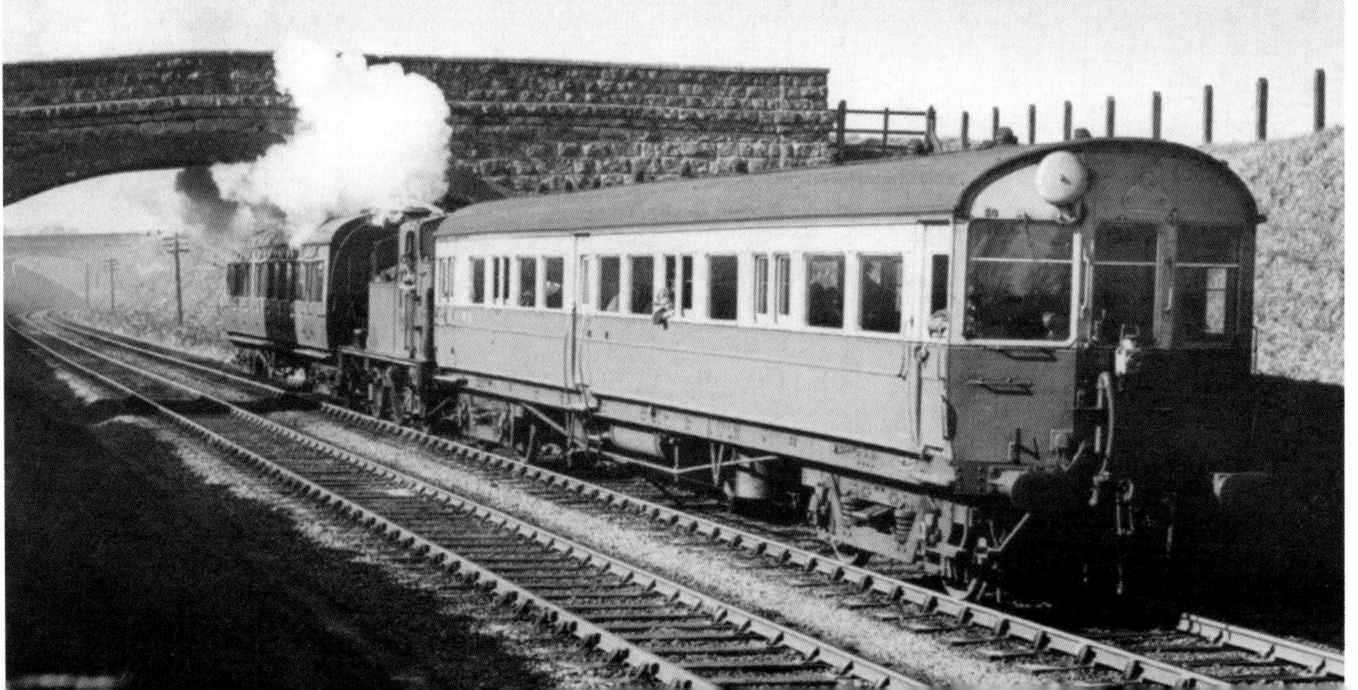

1412 crosses the road at Porthywaen with the 6.55pm Llangynog-Oswestry, 1 July 1950. (MLS Collection)

An unidentified 14XX sandwiched between the auto-trailer W50 and a corridor coach at Upton Scudamore between Westbury and Salisbury, c1953. I was evacuated with my mother here during the war and I remember being enticed as a 5-year-old by older boys into laying pennies on the line at Upton Scudamore crossing to be squashed by passing trains like this. (Andrew Wilson Collection)

Right and opposite above: **Three scenes** at Hemyock on 26 July 1954, with 1429 arriving at the terminus and then assembling the milk tanks for the return trip. Note that 1429 has acquired a large and conspicuous top feed behind the chimney. (MLS Collection)

Locomotives built by the GWR after 1923 • 115

A busy scene at Monmouth Troy with 1455 on the Ross-on-Wye auto train with a 64XX behind and a GW railcar of the later series in the other platform, 8 August 1951.
(MLS Collection)

1401, the engine used in the *Titfield Thunderbolt* film, with the auto train at Symond's Yat, near Tintern, c1955.
(MLS Collection)

1404 at Cinderford station in the Forest of Dean, 29 August 1952.
(MLS Collection)

Locomotives built by the GWR after 1923 • 117

1419 stands ready for departure with the 1.45pm to Lostwithiel at Fowey station, May 1960. (MLS Collection)

1428 in the platform at Oswestry with the auto train for Gobowen, 11 July 1951. (MLS Collection)

1424 slows ready to stop at Brimscombe Bridge Halt on a Gloucester-Chalford auto train, April 1959. (Michael Hale/GW Trust)

1432 arrives at Elson Halt with the 3.25pm Wrexham-Ellesmere, 16 March 1961. (MLS Collection)

Locomotives built by the GWR after 1923 • 119

1447 at Wallingford on the 12.47pm (SO) to Cholsey & Moulsford, 19 January 1957.
(MLS Collection)

1447 departs Radley for Abingdon with a tour organised by the Oxford University Railway Society, 23 March 1959.
(J.D. Edwards/GW Trust)

1471 at Penygraig on the last day of the passenger service to Llantrisant, 7 June 1958. This and other branch passenger services to Llantrisant had long gone before I became Area Manager Bridgend covering the former Tondu and Llantrisant depots in 1965. By then freight services were in the hands of the EE type 3 diesels (Cl 37).

1458 propelling the autocoach departing from Ellesmere for Wrexham, 28 April 1962. (MLS Collection)

Locomotives built by the GWR after 1923 • 121

1465 propelling the auto coach and hauling a van near Saltney Junction, 10 March 1958. (J. Peden/MLS Collection)

1462 marshals the milk tanks for the mixed train departure from Uffculme on the Hemyock branch, 10 August 1962. (MLS Collection)

Above: **1468 in** the platform at Uffculme with the Hemyock branch train formed of one ex-Barry Railway coach, 8 July 1959. (R.M. Casserley/MLS Collection)

Right: **Newton Abbot's** 1470 at Churston with the auto train for Brixham while 5976 *Ashwicke Hall* stands at the station with a Kingswear-Exeter stopping train, 11 June 1960. Note the coach/bus awaiting passengers off the train – competition with the 14XX and its coach? Shades of the *Titfield Thunderbolt* rivalry between train and local bus company. (MLS Collection)

The 48XX (later 14XX) could be found all over the GWR system and the next series of photographs will illustrate some of the locations where our photographers ventured to find them.

Unfortunately, I experienced personal travel and acquaintance with few of the locomotives described in this book. The day on Swansea Docks in August 1957 which I described earlier was one example. The rest were with the 14XX auto trains between 1957 and 1964. During cold November evenings in 1956 when I was attempting to gain a scholarship to one of the Oxford colleges (I failed), I whiled away the damp misty evenings on Oxford station and twice made forays out to Radley to connect with the auto train to Abingdon. Both times it was hauled by 1420 of Oxford and the journey was completed in total darkness and swirling fog and on each occasion I was the only occupant of the train, puzzling the guard on why I turned straight back from Abingdon on such a miserable night. But at least the train was warm, which was more than Oxford station was (or the college room where I was spending the nights).

In April 1961, armed with a 'priv ticket' from Paddington to Princes Risborough via Maidenhead and High Wycombe, I alighted from the suburban train hauled by 6169 at Bourne End and interrupted my journey with a trip to Marlow and back on the 'Marlow Donkey' (the nickname of the Bourne End-Marlow auto train). I got chatting to the driver and was invited into the cab of the auto coach on the outward journey and onto the

Above: **1420 departs** from Paddington station with a lightweight parcels train for Oxford, c1956. (MLS Collection)

Below: **Abingdon, the** station I scarcely saw, as I stayed in the warmth of the auto coach. Here 1447 has just arrived with the 6.10pm from Radley on a summer's evening, 20 June 1959. (MLS Collection)

1447 awaits departure time in Marlow's single platform with the 10.5am return trip of the 'Marlow Donkey' to Bourne End, 16 May 1960.
(MLS Collection)

My steed for my footplate trip on the 'Marlow Donkey' was 1453 seen exactly five years earlier (coincidentally on my birthday) a few years earlier at Weymouth, 16 May 1955. It is painted in the pre-1956 black livery that still adorned it in 1961. Presumably it must have had some light overhaul at the Works in the interim years without repainting.
(L.B. Lapper/F.K. Davies/John Hodge Collections)

1458 arriving at Ealing Broadway with the auto train for Greenford, June 1957. (David Maidment)

footplate of 1453 on the return. 1453 was quite high mileage and the ride was surprisingly rough, although according to the driver it was more the fault of the track than 1453. There was no opportunity for speed and a steady 30mph sufficed for timekeeping. I continued my journey from High Wycombe and then back to Marylebone behind LMR Fairburn 2-6-4Ts which had replaced the GC A5s and ER L1s by that time.

In 1957, whilst working temporarily at Old Oak Common during the gap between leaving school and starting a university course, I had two trips over the Ealing-Greenford branch. The first occasion was with King 4-6-0, 6024 *King Edward I*, running light engine on test after a hot box repair in Old Oak Common's Factory. The second, by contrast, was with 1458 and its pair of auto coaches, seen in the adjacent photograph running into Ealing Broadway station in June 1957.

I described earlier my somewhat belated attempt to experience the Chalford-Gloucester auto trains, one of the last services of this type still powered by 14XX locomotives and auto coach. I was appointed to my first permanent railway management post as stationmaster Aberbeeg, in April 1964 and still being single, used my half days off when not on call to escape from the valley and chase steam elsewhere. Trips to Pontypool Road, Hereford and Gloucester were most frequent and on one occasion I went beyond Gloucester on the 6.22pm to Chalford behind 1445 as described in the log I showed earlier. I kick myself now that I didn't take advantage of that opportunity more often and experience one of the high speed runs on that route which was fairly common knowledge among enthusiasts at that time.

The end came rapidly in the late 1950s/early 1960s as dieselisation with DMUs and branch line closures removed much of the work for which the 14XX were designed. The first withdrawals took place as early as February 1956 (1404, 1425 and 1460) and by the end of 1960 the class of seventy-five engines had been cut by over half to just thirty-four. The allocation of these was:

Gloucester:	7
Exeter:	5
Oxford:	4
Oswestry:	3
Slough:	2
Swindon:	2
Newton Abbot:	2
Laira:	2
Hereford:	2

1472 at Chalford with the auto train for Gloucester, c1964. 1472 is one of the engines reported to have exceeded 70mph between Stonehouse and Tuffley Junction on this service. (MLS Collection)

Three 14XX await their fate on the Swindon Dump, 12 April 1959. 1401 was withdrawn from Gloucester in November 1958 and 1423 from Oswestry in January 1959 having spent most of its career at Goodwick (Fishguard). (MLS Collection)

Locomotives built by the GWR after 1923 • 127

Aylesbury:	1
Southall:	1
Reading:	1
Bristol Bath Road:	1
St.Blazey:	1

The last survivors were the small clutch of engines based at Gloucester for the Chalford auto trains which finished in 1964 and 1442 and 1450 from Exmouth Junction shed, after the closure of Exeter (GW) in May 1965. 1442 was resuscitated to haul the last train to Tiverton in October 1965. Four have been preserved – 1420, 1442, 1450 and 1466 (see chapter 5). 1424, a long-term resident of Gloucester, is credited with the highest mileage of the class – 950,721 – whilst 1457, which was withdrawn from Machynlleth in early 1959, had the lowest at 413,967.

The end for 1469 being cut up at Caerphilly, 4 April 1959.
(MLS Collection)

The last part of the first 14XX withdrawn, 1404, having its frame cut up at Caerphilly, 5 April 1956.
(MLS Collection)

5800-5819, 0-4-2T, 1933

Twenty 0-4-2 tank engines identical to the 48XX were built in 1933 apart from the fitting of auto gear and ATC apparatus. 5801, 5803, 5809-5813, 5815, 5816 and 5819 were subsequently equipped with ATC. For dimensions, see the details of the 48XX on page 98. They were intended for branch freight and shunting work and seven went to the Bristol Division and five for the Worcester Division. An order was made for fifteen more, allocated 5820-5834, but it was cancelled and 1460-1474 built in their place.

At nationalisation, the Bristol Division had three each at Bath Road and Swindon. The three Swindon engines were used mainly on the Highworth branch and the local goods trains to Tetbury and Malmesbury. The Worcester Division had seven – at Worcester itself (2), Tetbury (1), Leominster (2) and Kington (2). Bala in the Wolverhampton Division had two and Wales had four at Pontrilas (1), Newcastle Emlyn (1), Brecon (1) and Llanfyllin (1). They were less useful than the auto-fitted 14XX and as branch lines closed, they became redundant earlier. Six of the class were actually in store at Swindon by 1954 (5806, 5807, 5812, 5814, 5817, 5819). Oxford in the London Division had gained 5803 and 5808 from the store at Swindon earlier. Eleven were withdrawn in 1957, starting with 5808 in February and including the six stored engines which were all withdrawn in June. The only one to survive into the 1960s was 5815, domiciled at Worcester for many years but finishing at Swindon. It achieved the highest mileage of the class, 637,990 miles. The lowest was 5814 with just 381,300.

There were no recorded logs of this class in the archives of the Railway Performance Society, but James Tawse in his *Steam World* article in February 1998 described a short journey on the Tetbury branch with Swindon's 5815 and a solitary Hawksworth non-corridor coach. Kemble was left three minutes late waiting connection off the 2.40pm Gloucester-Swindon but arrived in Tetbury on time including a stop at Rodmarton. The 7.2 miles took 15 minutes including the stop. The return journey, started 1½ minutes late, took 17½ minutes including stops at both Culkerton and Rodmarton and arrived at Kemble on time.

5806 shortly after construction in 1933. (MLS Collection)

Locomotives built by the GWR after 1923 • 129

5818 at Pontrilas shed, 24 August 1951. It looks well cared for by the local crews. (K. Cooper/ MLS Collection)

5811 at Swindon Works, 8 May 1955. (MLS Collection)

The prototype 5800, ex-works in the plain BR black livery that was applied to all the 58XX class. None were painted BR lined green like many of the 14XX.
(MLS Collection)

5801 at Builth Wells station with a train for Brecon, 29 March 1948.
(MLS Collection)

Locomotives built by the GWR after 1923 • 131

5801 at Barmouth with a train for Dolgelley, 23 June 1956. (F.M. Gates/GW Trust)

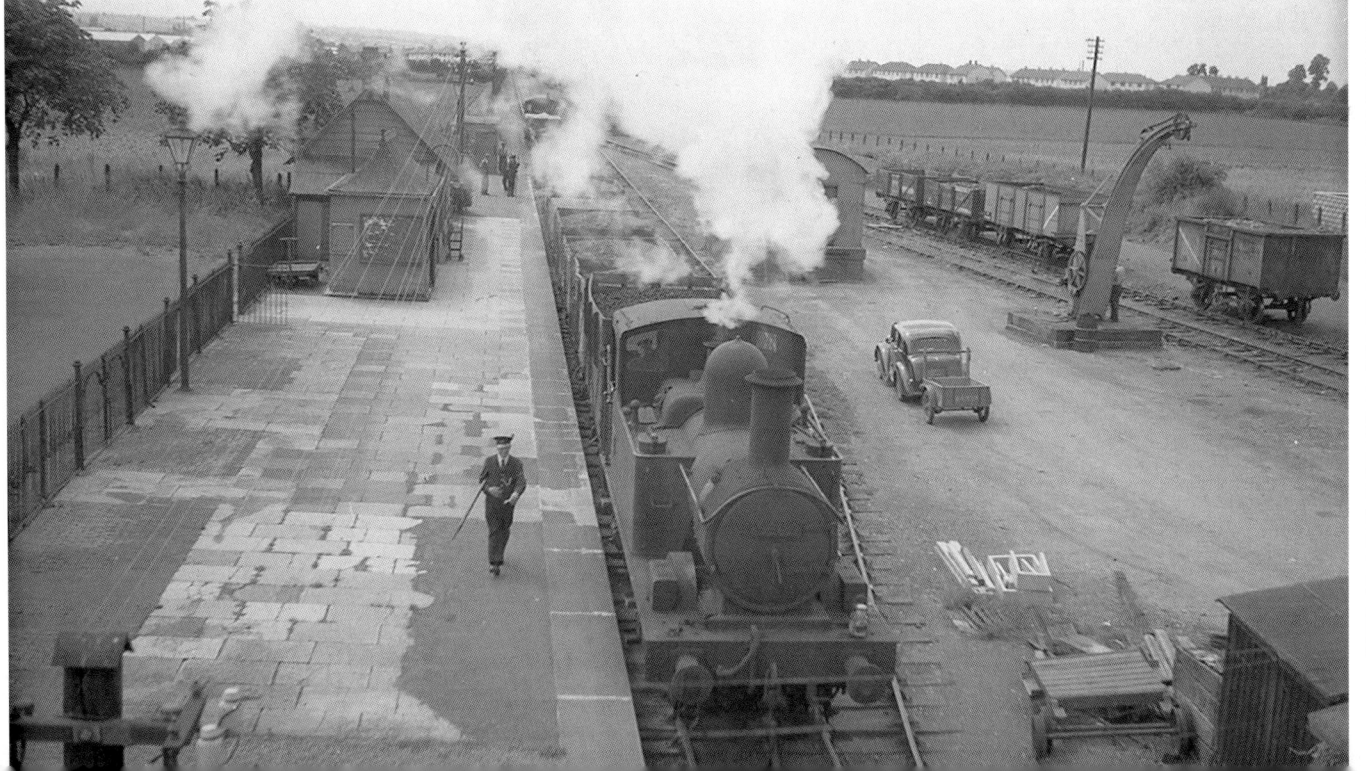

5805 at Highworth with a coal train for berthing for the local coal merchants, 21 July 1954. (Peter Fry/GW Trust)

132 • FOUR-COUPLED TANK LOCOMOTIVE CLASSES BUILT BY THE GREAT WESTERN RAILWAY

5800 of Swindon at Hannington on a Highworth train, August 1950. The withdrawal of passenger services from the branch took place in March 1953. (T.J. Edgington)

5802 of Swindon with the local goods, shunting at Malmesbury, 18 October 1952. (Millbrook House)

Locomotives built by the GWR after 1923 • 133

5810 at Blaenau Ffestiniog station with the 5.34pm train to Bala, 7 September 1957.
(MLS Collection)

5801 near Talyllyn with a Builth Wells – Brecon train, 25 June 1953.
(MLS Collection)

5810 takes water on the Bala-Blaenau Ffestiniog line, c1957. (MLS Collection)

Below left: **5815 runs** round its train at Tetbury before the return to Kemble, 28 June 1958. (H.D. Bowtell/MLS Collection)

Below right: **5815 at** Tetbury before the return to Kemble, on the journey described by James Tawse in the *Steam World* magazine, 10 May 1958. (James Tawse)

Locomotives built by the GWR after 1923 • 135

Above left: 5819 standing at Llandyssil station with the 1.30pm Newcastle Emlyn – Pencader train, 11 September 1952. (J.A. Peden/MLS Collection)

Above right: The 58XX were difficult to 'cop' for most trainspotters, especially those like me from the London area. The only one I ever saw was 5804 seen here at Swindon, on a rather dull misty day in April 1957. (David Maidment)

The end of 5815, the last survivor of the class, at the Swindon Dump awaiting scrapping, 28 April 1963. (MLS Collection)

Chapter 5
PRESERVATION

1420 in BR black livery at Swindon Stock Shed, 27 January 1952.
(MLS Collection)

1420 (GWR, 1933)
The Collett 0-4-2T 1420 was built in November 1933 and withdrawn in 1964. It was preserved and operated for a few years on the Dart Valley Railway, now renamed the South Devon Railway. It was painted in lined green with the letters G W R and the nameplate *Bulliver* underneath on the tank sides. It is currently cosmetically restored and is under overhaul for future operation.

1420 at Hereford in BR passenger lined green livery, 1 July 1962.
(A.W. Martin/MLS Collection)

1420 in operation on the South Devon Railway crossing the River Dart near Buckfastleigh. It is wearing the lettering 'Dart Valley' on the tank as the South Devon Railway was previously known. (Postcard/MLS Collection)

1420 operating on the South Devon Railway, seen here at Buckfastleigh, in lined green and adorned with the name *Bulliver*. (Author's Collection)

1420 preserved in late 1930s GWR livery at Ashburton, 1966.
(Peter Gray/GW Trust)

1420 piloted by preserved 4555 enter Worcester Shrub Hill with an SLS special railtour train, 19 September 1965.
(W. Potter)

1442 (GWR, 1935)

1442 was built in April 1935 and was one of the last two survivors, withdrawn from Exmouth Junction shed in May 1965. It had operated the Tiverton-Tiverton Junction branch (known locally as the 'Tivvy Bumper') and hauled the last train there in October 1965 (brought back specially from store awaiting breaking up). It was then retained as a static exhibit in Blundell's Road, Tiverton and is now on display in Tiverton Museum.

The 'Tivvy Bumper', 1442, on display at the Tiverton Museum. (Peter Gray/GW Trust)

1450 (GWR, 1935)

1450 was the other survivor at Exmouth Junction, also withdrawn in May 1965, after achieving 823,012 miles in traffic. It went to the Dart Valley Railway, as did 1420, but was purchased by Push-Pull Ltd in 2001 and moved to the Dean Forest Railway. It arrived on hire to the Severn Valley Railway in 2014 and has since been listed as part of the SVR home fleet. It was repainted in GWR livery in 2017. It was withdrawn from traffic on the SVR in December 2020 when its boiler certificate expired, after running 13,503 miles in service on that heritage line.

Above left: **1450** after repair and repainting at the Dean Forest workshops of Push-Pull Ltd, in the early part of the twenty-first century. (GW Trust)

Above right: **1450** and 1340 *Trojan* at the Didcot Great Western Railway Centre, 2019. (David Maidment)

1450 and 3717 (3440) *City of Truro* at an open day in Toddington on the Gloucestershire Warwickshire Railway, c2015. (David Maidment)

Preservation • 141

1450 and auto-trailer on the demonstration track at Didcot Railway Centre, 2017. (David Maidment)

1450 and restored auto-trailer 228 crossing the Nursery Pool Bridge during the South Devon Railway Gala, 27 April 1997. (Robin Stewart-Smith)

4866 as built – an official GWR photograph.
(F.K.Davies/John Hodge Collections)

1466 (GWR, 1936)

1466 was built in February 1936 and was based at Newton Abbot for most of its GWR/BR career. It was purchased from BR in 1964 for £750 and went to the Dart Valley Railway's Totnes depot. It moved to Didcot in 1967 and was in service there as 4866 in 1999 and 2000. It was repainted in BR black livery as 1466 for a while in 2013 but was later restored to GW green as 4866 and was anticipated to be back in operation in 2022 or 2023.

4866 repainted in GWR livery at Didcot Railway Centre, 2017.
(David Maidment)

1466 with preserved autocoach coming off the Wallingford branch at Cholsey on a special opening day, April 1968. (C.G. Stuart/GW Trust)

Isebrook at Burton Latimer, the former GW Sentinel No.12, that worked there from 1926-58, now at the Buckinghamshire Railway Centre. (F.K. Davies/John Hodge Collection)

12 *Isebrook*, Sentinel (GWR, 1926)

No.12 was a vertical-boilered Sentinel engine built by the Shrewsbury Sentinel Waggon Works in 1926, intended for branch passenger work with steam heating pipes and vacuum brakes. It was actually the second of two but the first was not taken into GW stock until trials had been completed. 12 was tested on the Fowey branch in the autumn of 1926 but was found unsatisfactory and returned to the makers, and was subsequently sold to Messrs T.E. Gray of Burton Latimer and named *Isebrook* and numbered 49. It was withdrawn in 1958 and was subsequently preserved and is at the Buckinghamshire Railway Centre, Quainton Road. For further details, see page 97.

GW Steam Railmotor No.93

Steam Railmotor No.93 was built at Swindon in 1908, one of a batch of thirty-five built there between 1905 and 1908. Initially allocated to Southall, it subsequently worked from Bristol, Croes Newydd, Gloucester, Taunton and Yatton. It was withdrawn from Swindon in 1934 and the coach was converted to auto trailer 212. This was condemned in 1956 but converted into a BR Work Study coach, acting as an office in Birmingham. It was acquired by the Great Western Society in 1970 and in 2011 it received a Heritage Lottery fund grant to be rebuilt as a steam driven working railmotor as built in 1908. It is currently operational at the Great Western Society's Centre at Didcot.

The restored railmotor No.93 in steam on the Didcot Railway Centre's short running line, c2017. (GW Trust)

COLOUR SECTION

Unfortunately, few colour photos exist of Great Western four-coupled tank engines except for the dainty 14XX which were found far and wide throughout the system. Laurence Waters of the Great Western Trust has made a good number available to me, the majority taken in the more picturesque parts of South Devon, the Exe Valley, the Hemyock, Morehampstead, Ashburton and Brixham branches.

1446 built in April 1935 and seen here in BR black livery at its home depot, Southall, September 1957. (C.G. Stuart/GW Trust)

146 • FOUR-COUPLED TANK LOCOMOTIVE CLASSES BUILT BY THE GREAT WESTERN RAILWAY

Above left: 1409 in unlined BR green livery as painted by Swindon Works, 13 April 1962. Seen here at its home depot, Gloucester. (C.G. Stuart/GW Trust)

Above right: 1445 in lined BR green passenger livery as painted at Swindon Works in the late 1950s, seen here at Marlow, 20 March 1962. (C.G. Stuart/GW Trust)

1453 departing from Marlow with the auto-train for Bourne End with trailer 220 *Wren*, 30 April 1960, a few months before the author rode in the trailer cab and the footplate of 1453 on the same turn. (T.B. Owen)

Above left: **1447 seen** at Hereford shortly before withdrawal, its lined green livery now fading, 1963. (C.G. Stuart/GW Trust)

Above right: **1445 had** just arrived at Marlow with the 'Marlow Donkey' – the push-pull train from Bourne End. This photo was taken on 20 March 1962. (C.G. Stuart/GW Trust)

A few minutes later 1445 rounds the curve into Bourne End with the 'Marlow Donkey' formed on this occasion with a single vehicle rather than auto coach, 20 March 1962. (C.G. Stuart/GW Trust)

Above left: **1444 waits** patiently in the bay platform at Radley with the auto-coach for Abingdon, 16 November 1959. (A.E. Doyle/GW Trust)

Above right: **1440 with** auto-trailer *Wren* at Princes Risborough with the branch train from Aylesbury, 1962. (M.E./Rail Online Trust)

A rather down-at-heel 1463 waits at Yatton with a local auto-coach working, probably for the Cheddar branch, October 1959. (Peter Gray/GW Trust)

Colour Section • 149

1456 with the Monmouth – Ross-on-Wye auto train nearing Symond's Yat beside the River Wye, 5 October 1957. (W. Potter/MLS Collection)

1455 propelling auto-trailer W237 is departing from Monmouth Troy for Ross-on-Wye, 28 May 1958. (T.B. Owen)

Gloucester's 1472 arriving at Chalford with the 11.20am Gloucester auto train, 12 September 1964. (W. Potter)

1453 near Stroud with the Gloucester - Chalford auto train, May 1964. It still bears the wooden hand painted numberplate on the smokebox door that it had three years earlier. (M. Hale/GW Trust)

Colour Section • 151

Gloucester's 1472 on a local auto train at Sharpness station, August 1960. (C.G. Stuart/GW Trust)

1470 approaching Ashburton with the branch train from Totnes, April 1958. (Peter Gray/GW Trust)

152 • FOUR-COUPLED TANK LOCOMOTIVE CLASSES BUILT BY THE GREAT WESTERN RAILWAY

Newton Abbot's 1466, now preserved, standing at Moretonhampstead after arriving with the branch train from Newton Abbot, February 1959. (Peter Gray/GW Trust)

1466 again taking its leisurely way to Moretonhampstead in February 1959. (Peter Gray/GW Trust)

Colour Section • 153

Another shot of 1466 at Tiverton in push-pull mode with an auto trailer, c1959. (Peter Gray/GW Trust)

An earlier shot of 1466 taking water at Torre station on the Newton Abbot-Kingswear line, December 1957. (Peter Gray/GW Trust)

Above left: Thirsty engines! This time it's 1470 taking water at Churston while operating the Brixham branch train, March 1960. (Peter Gray/GW Trust)

Above right: A different duty for Newton Abbot's faithful 1466 – a light pick-up goods at Exeter Cowley Bridge Junction making its way slowly towards Tiverton on the Exe Valley line, November 1961. (Peter Gray/GW Trust)

Right: **Now we see 1466** on the Hemyock branch marshalling a milk tank for the return trip to Tiverton Junction and Exeter. (Peter Gray/GW Trust)

Beside the still waters... 1451 and its Barry coach at Hemyock, September 1962. (Peter Gray/GW Trust)

1421 at Hemyock platform in November 1963, its old Barry coach replaced by an LNER Thompson vehicle, with the milk tanks in the background that will form a mixed train back to Tiverton Junction and Exeter. (Peter Gray/GW Trust)

1450 at Bampton station on the Exe Valley line, March 1963, shortly before the closure of the line. (Peter Gray/GW Trust)

1450 propels its auto-trailer beside the River Exe near Fairby, March 1963. (Peter Gray/GW Trust)

A busy scene at Tiverton Town station in August 1962 with three active 14xx, 1471, 1451 and 1462. 1471 has the auto-coach for Tiverton Junction and 1462 heads an Exe Valley train for Exeter. (Peter Gray/GW Trust)

A change of season – 1466 enters Tiverton Town station in the severe winter that afflicted the South West with unusually heavy snowfalls, February 1963. (Peter Gray/GW Trust)

Thorverton station, a passing place on the Exe Valley line. 1462 has just arrived with an afternoon train from Exeter whilst 1451 awaits the road propelling a Dulverton-Exeter service.
(Peter Gray/GW Trust)

1466 with its auto trailer at Tiverton , March 1962.
(Peter Gray/GW Trust)

A portrait of 1471 at rest on the Exe Valley branch, at Dulverton, 15 November 1962. (C.G. Stuart/GW Trust)

1471 with the LNER Thompson coach on the Hemyock branch train near Tiverton Junction, March 1963. (Peter Gray/GW Trust)

Above left: **1419 stands** at Fowey station with the branch train for Lostwithiel, c1960. (H.L./Rail Online Trust)

Above right: **1442 standing** at Exeter St David's after arrival with the Exe Valley branch train, 27 March 1963. (P dB/Rail Online Trust)

1473 at Chinnor with the *Six Counties Ltd* railtour, 3 April 1960. (C.R.R/Rail Online Trust)

APPENDIX

Included here are a selection of weight diagrams of the most important classes and statistics of building, rebuilding and withdrawal dates of all the classes described in the book. The locomotive dimensions are not repeated here but are included earlier in the appropriate chapter as indicated in the index.

Great Western Railway Broad Gauge
'Leo' Class 2-4-0T Broad Gauge
Statistics

Name	Built	Maker	Withdrawn
Elephant	1/1841	R&W Hawthorn	12/1870
Buffalo	3/1841	R&W Hawthorn	4/1865 stored, 7/1870
Dromedary	3/1841	R&W Hawthorn	12/1866 stored, 7/1870
Hecla	4/1841	Fenton, Murray & Jackson	9/1864 stored, 7/1870
Stromboli	4/1841	Fenton, Murray & Jackson	7/1870
Etna	6/1841	Fenton, Murray & Jackson	12/1870
Aries	6/1841	Rothwell	6/1871
Taurus	7/1841	Rothwell	12/1870
Gemini	9/1841	Rothwell	3/1866 stored, 7/1870
Cancer	10/1841	Rothwell	6/1874
Leo	10/1841	Rothwell	12/1870
Virgo	12/1841	Rothwell	12/1870
Libra	2/1842	Rothwell	6/1871
Scorpio	2/1842	Rothwell	12/1872
Sagittarius	4/1842	Rothwell	6/1871
Capricornus	4/1842	Rothwell	7/1870
Aquarius	6/1842	Rothwell	7/1870
Pisces	7/1842	Rothwell	6/1874

Corsair Class 4-4-0ST Broad Gauge Statistics

Name	Built	Maker	Withdrawn
Corsair	8/1849	Swindon	6/1873
Brigand	9/1849	Swindon	6/1873
Sappho	6/1854	R&W Hawthorn	12/1873
Homer	8/1854	R&W Hawthorn	12/1873
Virgil	9/1854	R&W Hawthorn	12/1873
Horave	9/1854	R&W Hawthorn	12/1880
Ovid	10/1854	R&W Hawthorn	3/1872
Juvenal	11/1854	R&W Hawthorn	12/1873
Seneca	11/1854	R&W Hawthorn	3/1872
Lucretius	12/1854	R&W Hawthorn	3/1872
Theocritus	12/1854	R&W Hawthorn	12/1873
Statius	1/1855	R&W Hawthorn	10/1871
Euripides	2/1855	R&W Hawthorn	12/1871
Hesiod	3/1855	R&W Hawthorn	2/1872
Lucan	3/1855	R&W Hawthorn	3/1872

'Metropolitan' 2-4-0 Condensing Tanks Broad Gauge Statistics

Name	Built	Maker	Withdrawn
Hornet*	6/1862	Vulcan Foundry	6/1873
Bee	7/1862	Vulcan Foundry	12/1874
Gnat	7/1862	Vulcan Foundry	6/1874
Wasp	8/1862	Vulcan Foundry	6/1877
Mosquito	8/1862	Vulcan Foundry	12/1875
Locust	8/1862	Vulcan Foundry	12/1876
Shah	6/1862	Kitson & Co.	6/1872
Bey	7/1862	Kitson & Co.	6/1872
Czar	8/1862	Kitson & Co.	6/1871
Mogul*	8/1862	Kitson & Co.	6/1872
Kaiser	9/1862	Kitson & Co.	6/1872
Khan	9/1862	Kitson & Co.	12/1872
Fleur de Lis	7/1863	Swindon	12/1872

Name	Built	Maker	Withdrawn
Rose	8/1863	Swindon	10/1877
Thistle	9/1863	Swindon	6/1874
Shamrock	11/1863	Swindon	12/1877
Camelia	12/1863	Swindon	6/1876
Azalia*	4/1864	Swindon	6/1872
Lily*	5/1864	Swindon	12/1872
Myrtle*	5/1864	Swindon	12/1873
Violet*	7/1864	Swindon	12/1872
Laurel*	10/1864	Swindon	6/1872

* Rebuilt as 2-4-0 tender locomotives

'Hawthorn' class rebuilt 2-4-0 Saddle Tanks Broad Gauge Statistics

Name	Built	Rebuilt	Maker	Withdrawn
Melling	5/1865	1877	Avonside Engine Co.*	5/1892
Roberts	6/1865	1877	Avonside Engine Co.	5/1892
Hedley	6/1865	1877	Avonside Engine Co.	5/1892**
Bury	7/1865	1877	Avonside Engine Co.	5/1892
Beyer	12/1865	1877	Avonside Engine Co.	4/1887
Penn	1/1866	1877	Avonside Engine Co.	5/1892
Stewart	1/1866	1877	Avonside Engine Co.	5/1892
Ostrich	12/1865	1877	Swindon	5/1892
Cerberus	2/1866	1877	Swindon	5/1892
Pollux	2/1866	1877	Swindon	5/1892

* Renamed from Slaughter, Gruning & Co. in 1865.
** Retained as stationary boiler, condemned 1914, scrapped 1929.

'3501' class 2-4-0 side tanks, 'Convertibles' Statistics

Number	Built	Rebuilt	Standard Gauge Tender
3501	3/1885	11/1890	
3502	3/1885	5/1890	
3503	3/1885		9/1892
3504	3/1885		6/1892
3505	4/1885	5/1890	

Number	Built	Rebuilt	Standard Gauge Tender
3506	4/1885		4/1892
3507	4/1885	5/1891	
3508	4/1885	5/1890	
3509	5/1885		9/1892
3510	5/1885		9/1892

'3541' class 0-4-2 saddle tanks, 'Convertibles' reb. 0-4-4 side tanks
Statistics

Number	Built	Rebuilt as 0-4-4T	To Std Gauge	Rebuilt as 4-4-0
3541	9/1888	4/1890	1/1892	10/1899
3542	9/1888	1/1891	11/1892	3/1899
3543	9/1888	7/1891	7/1891	9/1899
3544	10/1888	5/1890	7/1892	5/1900
3545	10/1888	4/1891	7/1892	8/1900
3546	10/1888	8/1890	8/1892	1/1900
3547	10/1888	6/1891	6/1891	1/1902
3548	11/1888	1/1891	11/1892	3/1901
3549	11/1888	4/1891	5/1892	8/1899
3550	11/1888	9/1888	8/1892	7/1900
3551	11/1888	7/1890	3/1892	12/1900
3552	12/1888	10/1890	8/1892	10/1900
3553	12/1888	5/1890	9/1892	1/1899
3554	1/1889	3/1891	3/1892	2/1900
3555	1/1889	11/1890	7/1892	4/1900
3556	2/1889	8/1890	2/1892	8/1901
3557	3/1889	2/1891	8/1892	11/1899
3558	6/1889	9/1890	5/1892	12/1901
3559	7/1889	10/1890	6/1892	1/1901
3560	7/1889	*	8/1892	9/1899

* Built as 0-4-4T

Great Western Railway Standard Gauge
Private company locomotives built for the GWR
91-92, Beyer, Peacock 0-4-2ST, 1857, reb.0-4-0ST 1893

Weight diagram

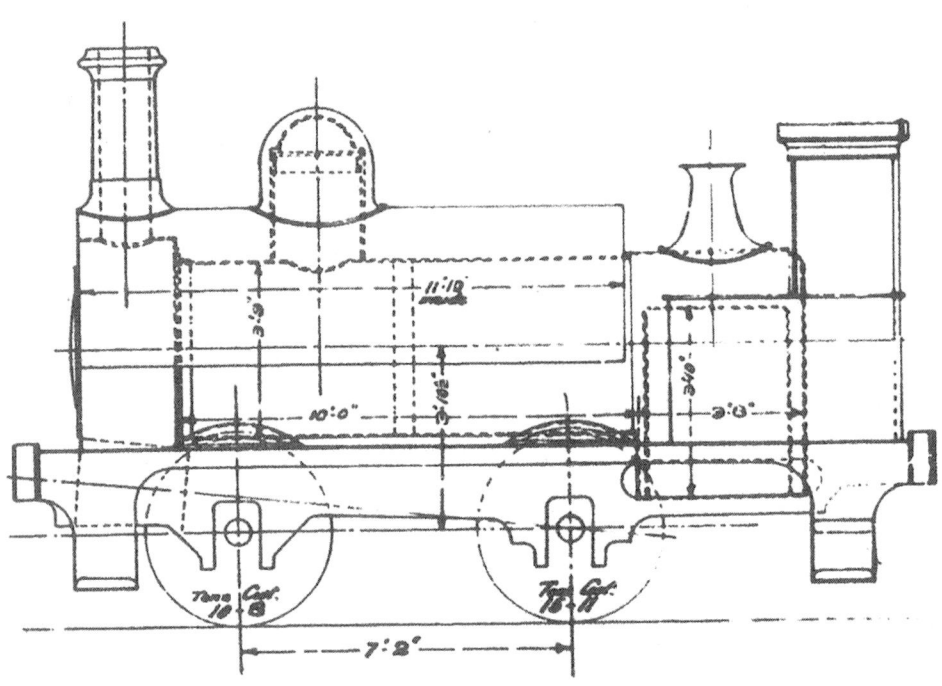

342, Beyer, Peacock 0-4-2ST, 1856, reb 0-4-0ST
343, Beyer, Peacock 2-4-0T, 1864
Statistics

No.	Built	Maker	Acquired	Rebuilt	Withdrawn
91	4/1857	Beyer, Peacock			1/1877
92	4/1857	Beyer, Peacock	12/1878 & 10/1893		7/1942
342	10/1856	Beyer, Peacock	2/1881 & 8/1897		8/1931
343	1865	Beyer, Peacock			1888

Great Western Wolverhampton built Locomotives
45, 0-4-0ST, 1880
Statistics

No.	Built	Maker	Withdrawn
45	6/1880	Wolverhampton	4/1938

2-4-0T 1864, reb.2-4-0ST, 1867
Statistics

No.	Built	Maker	Renumbered	Rebuilt	Withdrawn
1A	7/1864	Wolverhampton	17	10/1867	3/1889
2A	7/1864	Wolverhampton	18	6/1866	5/1892
3A	9/1864	Wolverhampton	1002	8/1867	12/1893
4A	10/1865	Wolverhampton	1003	4/1868	6/1886
11	11/1865	Wolverhampton		8/1867	1/1887
177	11/1865	Wolverhampton	227, 238	Unknown	4/1890
344	11/1865	Wolverhampton		5/1868	9/1893
345	12/1865	Wolverhampton		3/1868	11/1888
346	12/1865	Wolverhampton		2/1867	1/1888
347	1/1866	Wolverhampton		9/1866	5/1888
348	1/1866	Wolverhampton		12/1866	6/1883
349	1/1866	Wolverhampton		3/1867	8/1888

'517' class, 0-4-2T, 1868-1885
Weight diagram

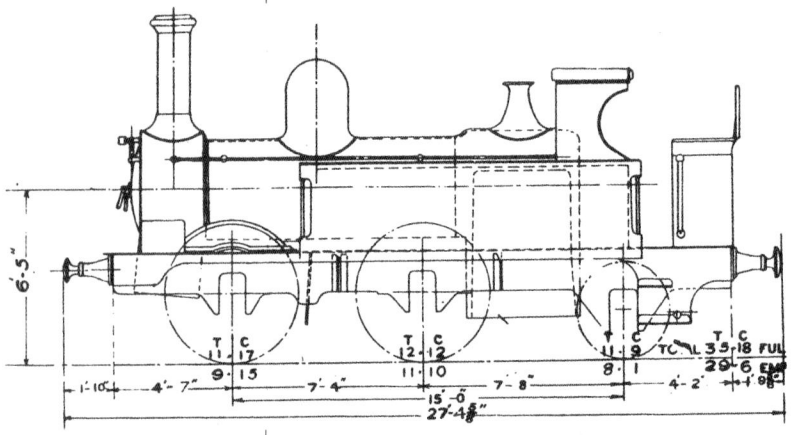
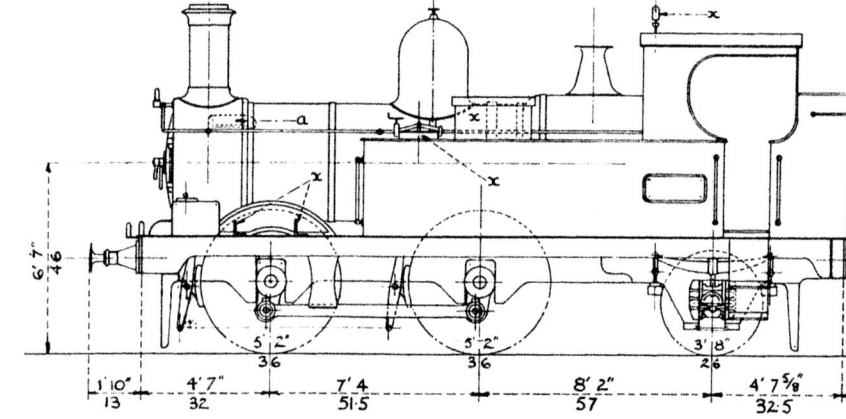

Statistics
All '517' class locomotives were built at Wolverhampton Stafford Road Works

No.	Built	Renumbered	Rebuilt (to long wheelbase) auto- fitted	Withdrawn
1040	4/1868	517	7/1900	2/1934
1041	5/1868	518	12/1896	2/1928
1042	5/1868	519		4/1933

No.	Built	Renumbered	Rebuilt (to long wheelbase) auto- fitted	Withdrawn
1043	5/1868	520	6/1903	10/1912
1044	6/1868	521	9/1915	4/1930
1045	6/1868	522		10/1935
1046	7/1868	523	12/1907	7/1934
1047	7/1868	524	4/1886	10/1930
1048	7/1868	525	3/1914	1/1934
1049	8/1868	526	10/1895	9/1933
1050	8/1868	527	5/1902	9/1928
1051	9/1868	528	5/1885	10/1935
1052	9/1868	529	12/1884	9/1930
1053	10/1868	530		5/1932
1054	10/1868	531	2/1901	10/1929
1055	10/1868	532		12/1929
1056	11/1868	533	9/1884	1/1934
1057	11/1868	534	9/1885	7/1932
1058	12/1868	535	4/1886	3/1929
1059	12/1868	536	3/1904	9/1929
1060	12/1868	537	10/1894	10/1932
1061	1/1869	538		12/1934
1062	1/1869	539	4/1910	5/1933
1063	1/1869	540	5/1884	5/1933
1064	2/1869	541	1/1909	10/1929
1065	2/1869	542	5/1899	4/1934
1066	3/1869	543	9/1884	3/1929
1067	3/1869	544	4/1899	10/1928
1068	4/1869	545		9/1928
1069	4/1869	546	4/1911	9/1928
1070	4/1869	547	10/1911	5/1930
1071	5/1869	548	4/1908	10/1934
1072	5/1869	549	2/1895	8/1929
1073	6/1869	550	9/1902	1/1929
1074	6/1869	551	9/1895	8/1929
1075	6/1869	552	6/1894	7/1915
1076	7/1869	553	7/1908	6/1932
1077	7/1869	554	9/1900	5/1933

No.	Built	Renumbered	Rebuilt (to long wheelbase) auto- fitted	Withdrawn
1078	8/1869	555	4/1902	7/1930
1079	8/1869	556	4/1911	9/1933
1080	8/1869	557		5/1933
1081	9/1869	558	10/1885	9/1934
1082	9/1869	559		2/1934
1083	10/1869	560		2/1913
1084	10/1869	561	10/1884	2/1929
1085	10/1869	562	2/1903	4/1929
1086	11/1869	563	12/1883	6/1929
1087	11/1869	564	1/1900	11/1933
1100	11/1869	565	10/1885	12/1930
1101	12/1869	566	7/1900	5/1934
1102	12/1869	567		1905 (Sold)
1103	12/1869	568	6/1885	3/1932
1104	1/1870	569	10/1895	5/1933
1105	1/1870	570	5/1899	10/1934
1106	1/1870	571	3/1896	2/1936
1107	3/1870	572	12/1894	9/1928
1108	3/1870	573	9/1895	2/1915
1109	4/1870	574	4/1897	2/1934
1110	4/1870	575	1/1895	12/1931
1111	5/1870	576	1/1897	7/1931
826	9/1873		4/1901	10/1928
827	9/1873			5/1933
828	10/1873		8/1901	11/1934
829	10/1873		3/1900	3/1934
830	10/1873		10/1895	1/1936
831	11/1873		7/1897	12/1935
832	11/1873		3/1897	4/1930
833	11/1873		10/1895	3/1934
834	12/1873			5/1933
835	12/1873			2/1935
836	1/1874		9/1902	12/1930
837	1/1874		2/1911	2/1934
838	2/1874		11/1897	12/1929
839	3/1874		8/1897	6/1932

No.	Built	Renumbered	Rebuilt (to long wheelbase) auto- fitted	Withdrawn
840	3/1874		6/1903	8/1929
841	4/1874		6/1896	2/1931
842	4/1874			9/1930
843	5/1874		8/1896	7/1915
844	11/1874		2/1897	3/1913
845	12/1874		6/1902	12/1934
846	1/1875		6/1896	12/1932
847	2/1875		6/1896	5/1933
848	2/1875		10/1899	6/1945
849	3/1875		4/1900	4/1904
1154	10/1875		3/1913	11/1934
1155	11/1875		1/1897	2/1936
1156	11/1875		11/1898	10/1930
1157	12/1875		12/1896	11/1935
1158	12/1875		2/1900	3/1931
1159	1/1876		3/1896	8/1947
1160	1/1876		5/1901	10/1932
1161	2/1876		10/1895	10/1945
1162	2/1876		5/1907	12/1934
1163	3/1876		8/1895	5/1946
1164	3/1876		11/1906	12/1937
1165	3/1876		12/1894	3/1930
202	4/1876		9/1895	9/1928
203	5/1876			4/1920
204	5/1876			11/1904
205	5/1876		1/1897	1/1934
215	6/1876		1/1910	1/1934
216	6/1876		2/1898	12/1936
217	6/1876		5/1898	2/1934
218	7/1876			9/1930
219	7/1876		3/1911	8/1934
220	7/1876		10/1899	12/1929
221	8/1876		2/1898	9/1929
222	8/1876		7/1903	11/1928
1421	4/1877		1/1901	9/1928
1422	5/1877		6/1896	7/1928

No.	Built	Renumbered	Rebuilt (to long wheelbase) auto- fitted	Withdrawn
1423	5/1877		8/1899	2/1930
1424	5/1877		11/1895	9/1933
1425	6/1877		5/1898	8/1932
1426	6/1877			11/1934
1427	7/1877			6/1837
1428	7/1877		3/1899	10/1932
1429	7/1877			10/1932
1430	8/1877		10/1898	10/1936
1431	8/1877			1/1933
1432	9/1877		2/1900	10/1932
1433	10/1877		8/1897	4/1936
1434	10/1877		5/1896	8/1928
1435	11/1877		9/1904	9/1932
1436	11/1877			11/1944
1437	11/1877		10/1900	9/1928
1438	11/1877			8/1934
1439	12/1877		11/1907	9/1930
1440	12/1877		8/1901	9/1935
1441	1/1878		4/1899	8/1929
1442	1/1878		8/1899	5/1945
1443	1/1878		10/1908	12/1934
1444	2/1878		10/1899	11/1934
1465	1/1883		12/1899	1/1936
1466	1/1883		8/1908	6/1935
1467	2/1883		3/1908	3/1929
1468	2/1883		2/1903	11/1936
1469	3/1883		1/1902	2/1931
1470	4/1883		8/1897	12/1934
1471	4/1883		11/1900	4/1932
1472	5/1883		3/1913	11/1934
1473 *Fair Rosamund*	5/1883			8/1935
1474	6/1883		6/1901	4/1931
1475	7/1883		2/1900	8/1932
1476	7/1883		10/1900	6/1929
1477	10/1884		12/1894	12/1937

No.	Built	Renumbered	Rebuilt (to long wheelbase) auto- fitted	Withdrawn
1478	10/1884		4/1904	5/1936
1479	11/1884		10/1901	4/1932
1480	11/1884			6/1913
1481	12/1884			2/1930
1482	1/1885		1/1900	6/1935
1483	6/1885		As built	8/1932
1484	7/1885		As built	5/1930
1485	8/1885		As built	12/1936
1486	9/1885		As built	1/1936
1487	10/1885		As built	8/1936
1488	11/1885		As built	11/1937

3571-3580, 0-4-2T, 1895
Weight diagram

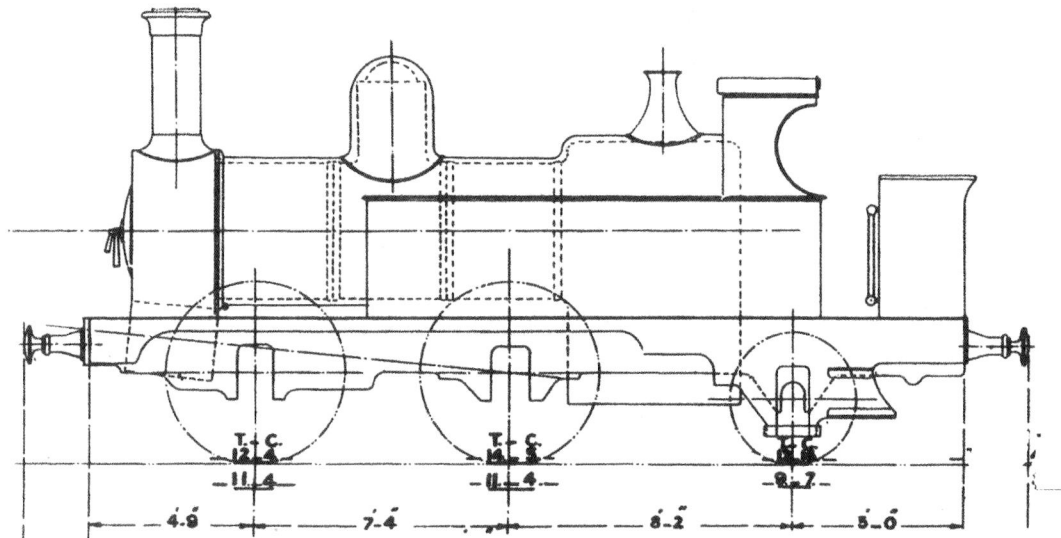

Statistics
All '3571' class locomotives were built at Wolverhampton Stafford Road Works

No.	Built	Withdrawn
3571	5/1895	10/1942
3572	6/1895	10/1928
3573	7/1895	12/1945
3574	8/1895	12/1949
3575	9/1895	10/1949

No.	Built			Withdrawn
3576	9/1895			6/1929
3577	10/1895			5/1949
3578	11/1895			7/1945
3579	12/1895			9/1942
3580	1/1897			12/1945

Swindon designed classes
117, 4-4-0T, 1854
320-1, 2-4-0WT, 1864
Statistics

No.	Built	Maker	Rebuilt	Withdrawn
117	6/1854	Swindon		1863
320	10/1864	Swindon	8/1867	7/1881
321	10/1864	Swindon	1/1873	7/1881

The '455' class 'Metro Tank' 2-4-0T, 1869-1899
Weight diagrams

The 1874 longer wheelbase design with later Collett enclosed cab and bunker.

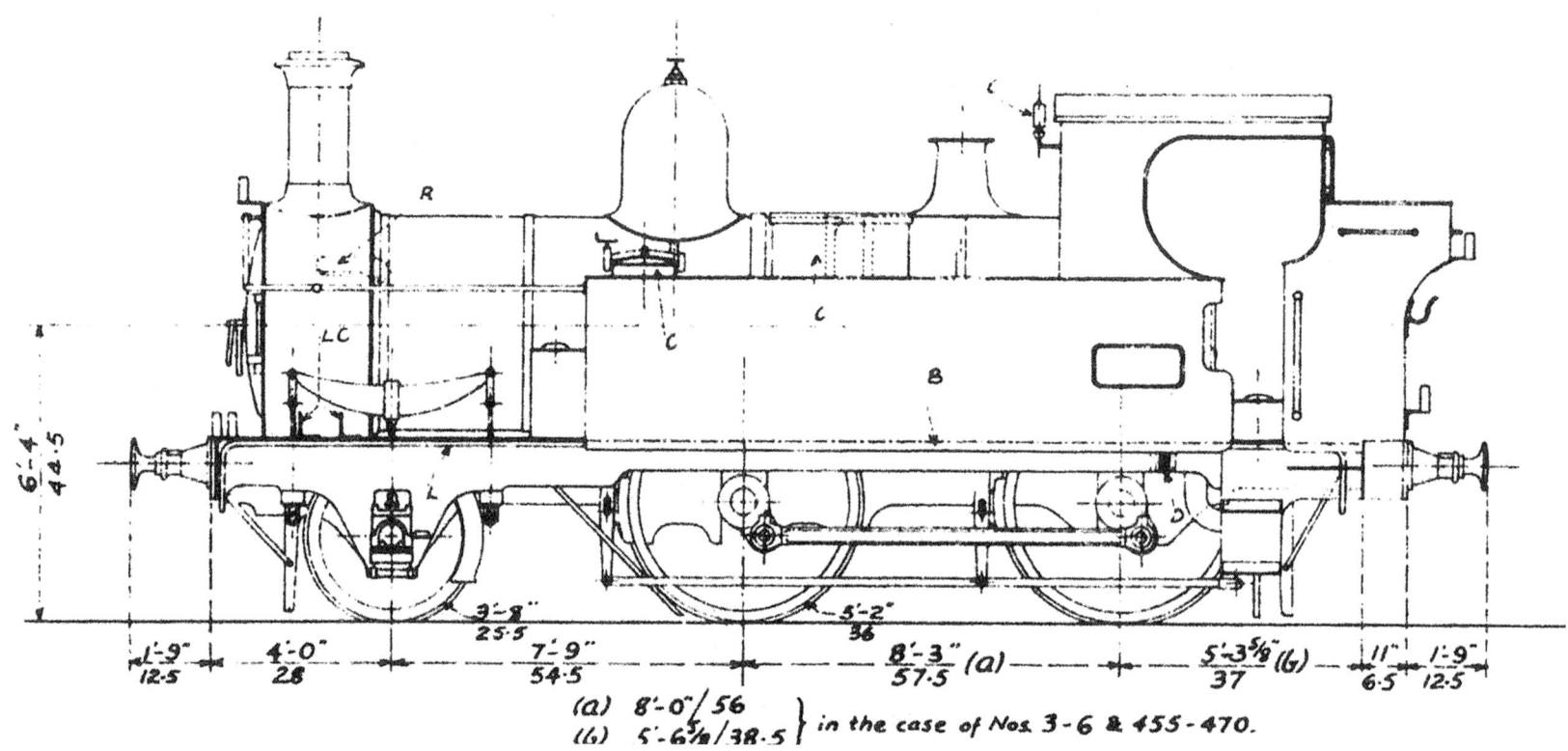

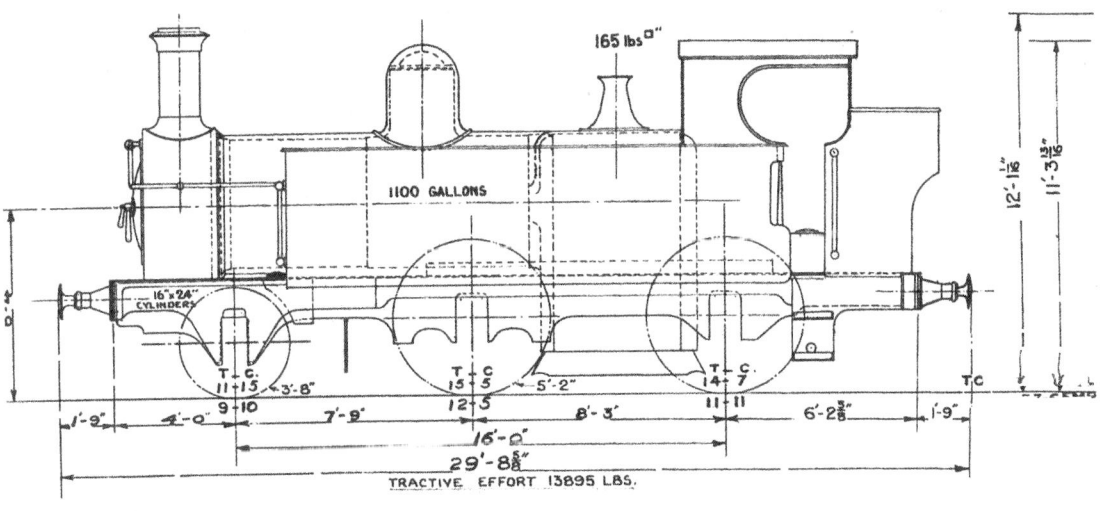

The 1894 1,100 gallon tank design with later Collett enclosed cab and bunker.

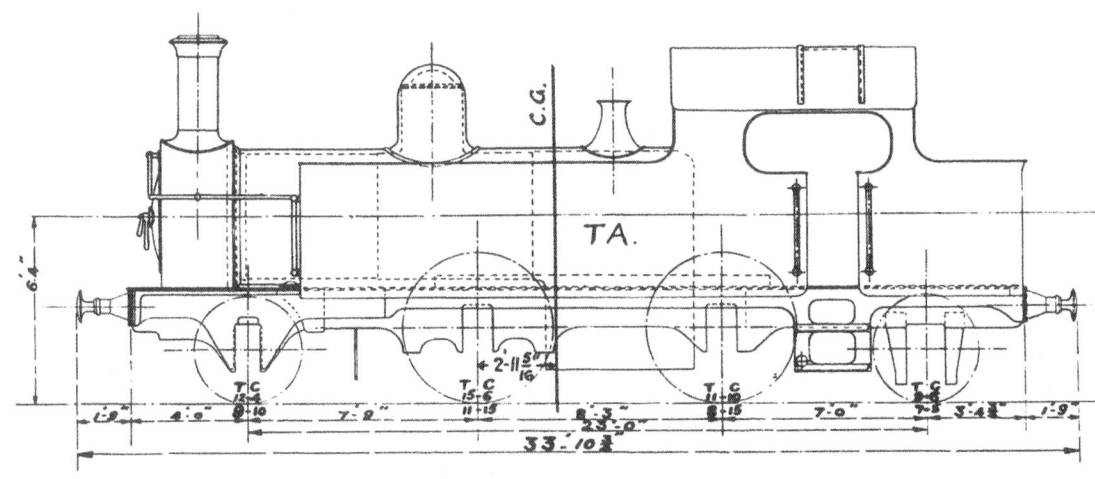

The 1899 'large' Met Tank rebuilt in 1906 as a 2-4-2T.

Statistics

No.	Built	Condensing * removed	Large Tanks Later gallons	Auto	ATC * & trip cocks	Closed cab	Withdrawn
455	1/1869	Yes *	840				8/1906
456	1/1869	Yes * 6/83	840			Yes	1/1931
457	1/1869	Yes * 5/83	840		*Yes	Yes	11/1934
458	2/1869	Yes * 8/86	840		*Yes	Yes	4/1933
459	2/1869	Yes * 2/80	840	12/28	Yes		10/1933
460	2/1869	Yes * 9/83	840		Yes	Yes	5/1932
461	2/1869	Yes * 6/85	840			Yes	12/1930
462	2/1869	Yes * 2/83	840				3/1907

No.	Built	Condensing * removed	Large Tanks Later gallons	Auto	ATC * & trip cocks	Closed cab	Withdrawn
463	3/1869	Yes *	840			Yes	6/1929
464	3/1869	Yes *	840		Yes		12/1934
465	3/1869	Yes * 9/83	840			Yes	2/1931
466	3/1869	Yes *	840				4/1907
467	4/1869	Yes *	840				12/1930
468	4/1869	Yes *	840	5/29	Yes	Yes	2/1932
469	4/1869	Yes *	840			Yes	9/1928
470	4/1869	Yes *	840	5/28	Yes	Yes	2/1934
3	4/1869	Yes * 4/84					9/1906
4	4/1869	Yes *	850				3/1913
5	5/1869	Yes *					10/1932
6	5/1869	Yes *	860				10/1928
613	7/1871	Yes *	820			Yes	2/1931
614	7/1871	Yes * 3/80	820				7/1906
615	8/1871	Yes *	820	7/30	Yes	Yes	2/1931
616	8/1871	Yes *	820		Yes	Yes	2/1931
617	8/1871	Yes *	820	10/30	Yes	Yes	5/1934
618	8/1871	Yes *	820				3/1907
619	8/1871	Yes *	820				1/1906
620	8/1871	Yes *	820			Yes	11/1928
621	9/1871	Yes *	820			Yes	4/1931
622	9/1871	Yes *	820				4/1898
623	9/1871		820	3/28	Yes	Yes	10/1930
624	9/1871		820	6/28			7/1929
625	9/1871		820				6/1903
626	9/1871		820		Yes	Yes	12/1932
627	10/1871		820				10/1928
628	10/1871		820				3/1930
629	10/1871		820				12/1912
630	10/1871		820				1/1929
631	10/1871		820				3/1898
632	10/1871		820				11/1929
967	8/1874	Yes *	860	12/29	Yes	Yes	4/1934
968	8/1874	Yes *	860				10/1906

No.	Built	Condensing * removed	Large Tanks Later gallons	Auto	ATC * & trip cocks	Closed cab	Withdrawn
969	9/1874	Yes *	860				8/1907
970	9/1874	Yes *	860				5/1904
971	9/1874	Yes *	860				11/1934
972	9/1874	Yes *	860		Yes	Yes	2/1932
973	9/1874	Yes *	860		Yes		5/1932
974	9/1874	Yes *	860				11/1904
975	9/1874	Yes *	860	8/29	Yes	Yes	4/1934
976	10/1874	Yes *	860		*Yes	Yes	11/1930
977	10/1874		860			Yes	3/1931
978	10/1874		860				8/1907
979	10/1874		860		Yes		9/1933
980	10/1874		860	3/29	Yes	Yes	8/1933
981	10/1874		860			Yes	7/1930
982	10/1874		860		*Yes	Yes	12/1933
983	11/1874		860	5/29	Yes	Yes	3/1934
984	11/1874		860				9/1906
985	11/1874		860				12/1930
986	11/1874		860		Yes	Yes	3/1934
1401	5/1878	Yes *	1080		Yes		1/1934
1402	5/1878	Yes *	1080				9/1906
1403	5/1878	Yes *	1080				11/1930
1404	6/1878	Yes *	1080		*Yes	Yes	2/1934
1405	6/1878	Yes *	1080				6/1929
1406	6/1878	Yes *	1080			Yes	6/1929
1407	6/1878	Yes *	1080		*Yes		4/1931
1408	7/1878	Yes *	1080	12/29	Yes	Yes	12/1934
1409	7/1878	Yes *	1080			Yes	7/1930
1410	7/1878	Yes *	1080		Yes		11/1934
1411	8/1878		1080				7/1929
1412	8/1878		1080				11/1906
1413	8/1878		1080		Yes	Yes	3/1931
1414	9/1878		1080				1/1929
1415	9/1878		1080	1/30	*Yes	Yes	4/1938
1416	9/1878		1080		*Yes	Yes	11/1930

No.	Built	Condensing * removed	Large Tanks Later gallons	Auto	ATC * & trip cocks	Closed cab	Withdrawn
1417	9/1878		1080		*Yes	Yes	4/1928
1418	10/1878		1080				4/1929
1419	10/1878		1080	10/30	Yes	Yes	7/1932
1420	10/1878		1080	1/30	*Yes	Yes	4/1936
1445	9/1881		840	6/29		Yes	12/1937
1446	10/1881		1080	6/28	Yes	Yes	2/1936
1447	10/1881		840			Yes	11/1928
1448	11/1881		840				7/1929
1449	11/1881		840				9/1928
1450	12/1881		840		Yes		4/1928
1451	12/1881		840		Yes		9/1928
1452	12/1881		840				9/1906
1453	1/1882		840	7/28	Yes	Yes	7/1933
1454	2/1882		840		Yes		11/1934
1455	2/1882		840	11/28	Yes	Yes	6/1935
1456	2/1882		840	7/28		Yes	10/1930
1457	3/1882		840				2/1929
1458	3/1882		840				4/1919
1459	3/1882		840	7/28	Yes	Yes	3/1936
1460	4/1882		1080		Yes		4/1913
1461	4/1882		840		Yes	Yes	8/1930
1462	4/1882		1080				3/1929
1463	5/1882		840	12/29	Yes	Yes	12/1934
1464	5/1882		840	4/28	Yes	Yes	1/1936
1491	6/1892		840		Yes		3/1930
1492	6/1882		840		Yes		2/1935
1493	7/1892		840	12/28	Yes	Yes	5/1935
1494	7/1892		840	5/29	Yes	Yes	12/1937
1495	8/1892		840	10/29	Yes	Yes	12/1938
1496	8/1892		840		Yes	Yes	10/1935
1497	8/1892		840		Yes		6/1938
1498	8/1892		840	12/28	Yes	Yes	9/1944
1499	9/1892		840	5/30	Yes	Yes	5/1946
1500	9/1892		840	6/28	Yes	Yes	9/1937
3561	1/1894	Yes *	1080	7/28	Yes	Yes	10/1949

No.	Built	Condensing * removed	Large Tanks Later gallons	Auto	ATC * & trip cocks	Closed cab	Withdrawn
3562	2/1894	Yes *	1080		Yes	Yes	2/1949
3563	2/1894	Yes *	1080		Yes	Yes	12/1944
3564	3/1894	Yes *	1080		Yes		1/1944
3565	3/1894	Yes *	1080		Yes	Yes	6/1936
3566	3/1894	Yes *	1080		Yes		4/1938
3567	4/1894	Yes	1080		Yes	Yes	9/1936
3568	4/1894	Yes	1080		Yes	Yes	11/1945
3569	4/1894	Yes *	1080		Yes		3/1936
3570	4/1894	Yes *	1080		Yes	Yes	12/1938
3581	1/1899	Yes *	1100	7/25	Yes	Yes	11/1945
3582	2/1899	Yes *	1100	7/25	Yes	Yes	11/1949
3583	2/1899	Yes *	1100		Yes	Yes	12/1947
3584	4/1899	Yes *	1100	10/29	Yes	Yes	11/1945
3585	5/1899	Yes *	1100		Yes		1/1948
3586	6/1899	Yes	1100	6/34	Yes		11/1949
3587	8/1899		1100	9/28	Yes	Yes	12/1944
3588	8/1899		1100		Yes	Yes	12/1949
3589	9/1899		1100		Yes		8/1848
3590	10/1899		1100	7/25	Yes	Yes	4/1944
3591	9/1899	Yes	1100		Yes		5/1936
3592	9/1899	Yes	1100		Yes		4/1949
3593	9/1899	Yes *	1100			Yes**	11/1927 Reb 2-4-2T 1906
3594	10/1899	Yes *	1100	10/29	Yes	Yes	5/1947
3595	10/1899	Yes *	1100	12/32	Yes	Yes	1/1945
3596	10/1899	Yes *	1100		Yes	Yes**	3/1945
3597	10/1899	Yes *	1100	12/28	Yes	Yes	8/1948
3598	10/1899	Yes *	1100		Yes	Yes	10/1936
3599	11/1899	Yes *	1100	11/29	Yes	Yes	10/1949
3600	11/1899	Yes *	1100		Yes	Yes	4/1942 renumbered 3500 1912

* Condensing gear removed after 1907
** High vaulted cab roof

3511-3520, 2-4-0T, 1885
Statistics

No	Built	Rebuilt as tender 'Stella' class
3511	5/1885	1/1894
3512	5/1885	3/1895
3513	6/1885	12/1894
3514	6/1885	11/1895
3515	7/1885	9/1894
3516	7/1885	9/1894
3517	7/1885	4/1894
3518	8/1885	11/1894
3519	8/1885	10/1894
3520	9/1885	1/1894

3521-3540, 0-4-2T, 1887, reb 0-4-4T, 1891
Statistics

No	Built	Rebuilt as 0-4-4T	Rebuilt as 4-4-0
3521	8/1887	10/1891	8/1899
3522	8/1887	2/1892	5/1899
3523	8/1887	4/1892	10/1899
3524	9/1887	1/1892	3/1902
3525	9/1887	1/1892	3/1901
3526	10/1887	4/1892	6/1900
3527	10/1887	4/1892	2/1900
3528	11/1887	4/1892	6/1900
3529	11/1887	4/1892	4/1900
3530	11/1887	5/1892	11/1899
3531	11/1887	11/1891	6/1901
3532	12/1887	3/1892	10/1902
3533	12/1887	3/1892	6/1901
3534	12/1887	3/1892	12/1899
3535	1/1888	3/1892	8/1899
3536	1/1888	12/1891	10/1901
3537	2/1888	3/1892	5/1899
3538	2/1888	2/1892	10/1900
3539	3/1888	4/1892	1/1900
3540	3/1888	4/1892	9/1900

3600—3630, 2-4-2T, 1900
Weight diagram

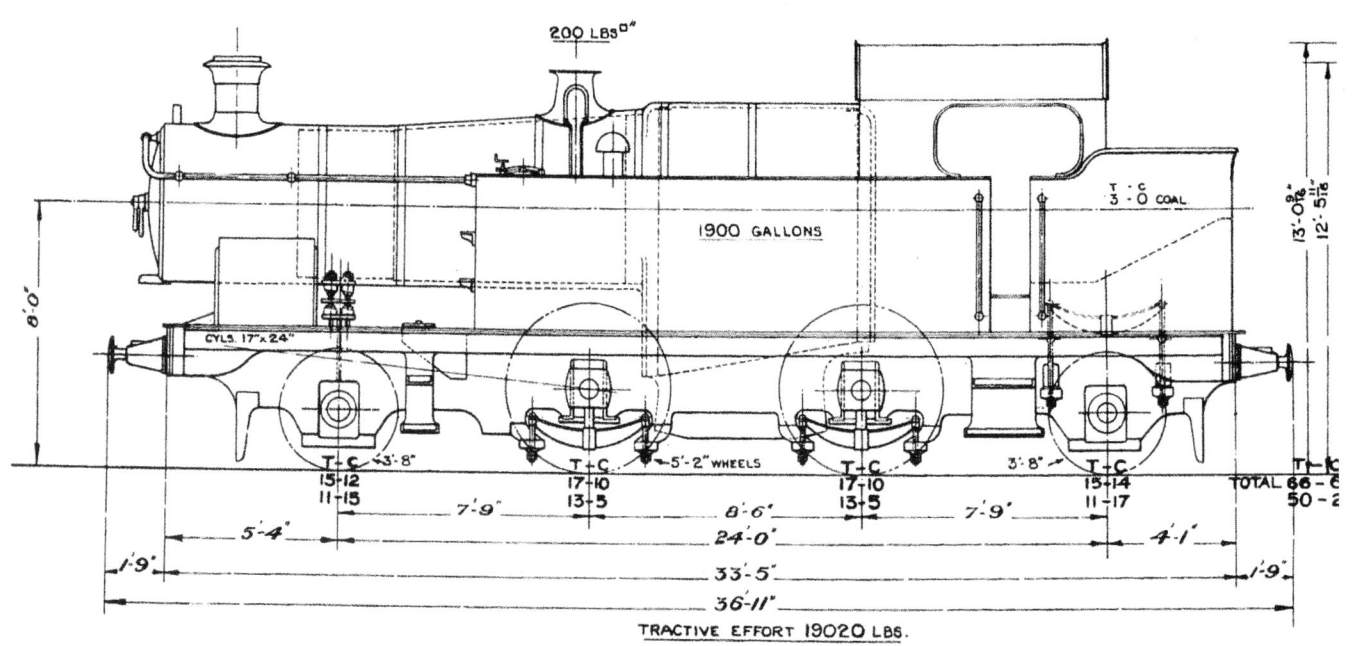

Front & rear view

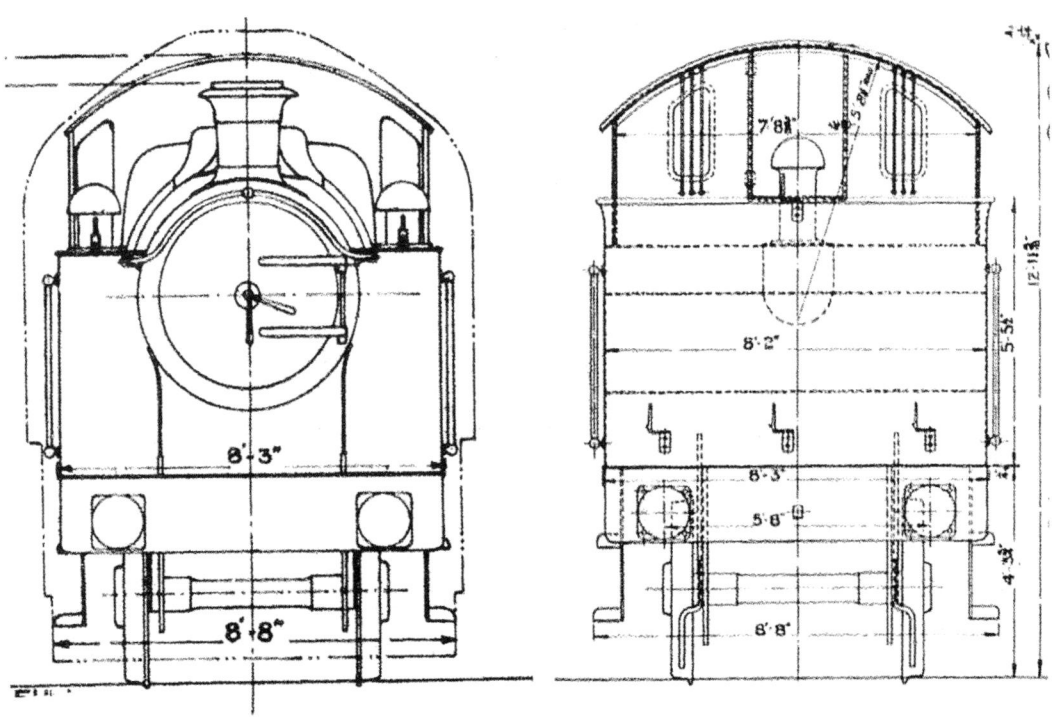

Statistics

No	Built	Superheated	Withdrawn	
11	12/1900	6/1911	6/1933	Renumbered 3600 in 1912
3601	2/1902	5/1915	3/1931	
3602	2/1902	12/1922	2/1931	
3603	3/1902	9/1924	/1931	
3604	3/1902	7/1917	11/1934	
3605	4/1902	5/1922	12/1930	
3606	4/1902	10/1912	3/1931	
3607	5/1902	10/1921	4/1931	
3608	5/1902	1/1920	4/1931	
3609	5/1902	1/1923	3/1932	
3610	5/1902	9/1924	11/1934	
3611	6/1902	2/1927	2/1931	
3612	6/1902	2/1926	10/1930	
3613	6/1902	6/1917	5/1932	
3614	6/1902	2/1913	4/1931	
3615	7/1902	5/1912	4/1931	
3616	7/1907	10/1921	9/1933	
3617	7/1902	6/1922	3/1933	
3618	7/1902	10/1912	/1934	
3619	8/1902	10/1913	4/1931	
3620	8/1902	3/1923	5/1931	
3621	10/1903	3/1927	1/1933	
3622	10/1903	3/1924	5/1932	
3623	11/1903	3/1925	4/1931	
3624	11/1903	2/1918	10/1931	
3625	11/1903	9/1922	7/1931	
3626	12/1903	3/1922	8/1932	
3627	12/1903	6/1923	1/1934	
3628	12/1903	1/1920	11/1934	
3629	12/1903	6/1912	11/1931	
3630	12/1903	10/1915	10/1930	

2221 class 4-4-2T (County Tanks)
Weight Diagrams
As built

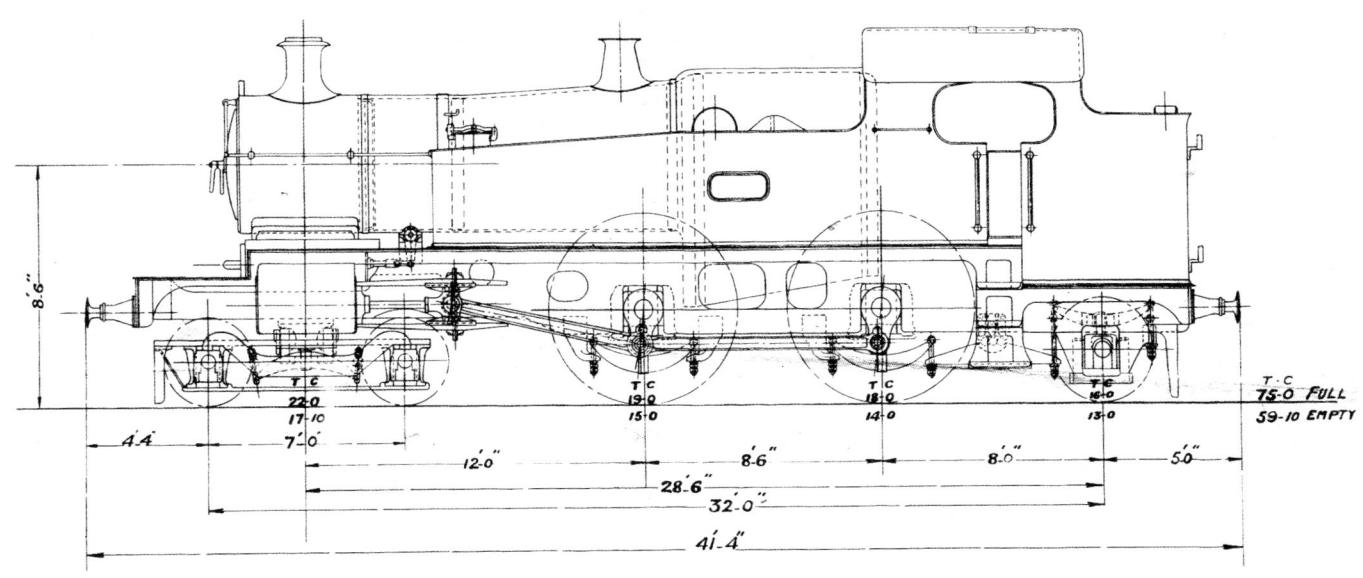

With extended smokebox and larger bunker

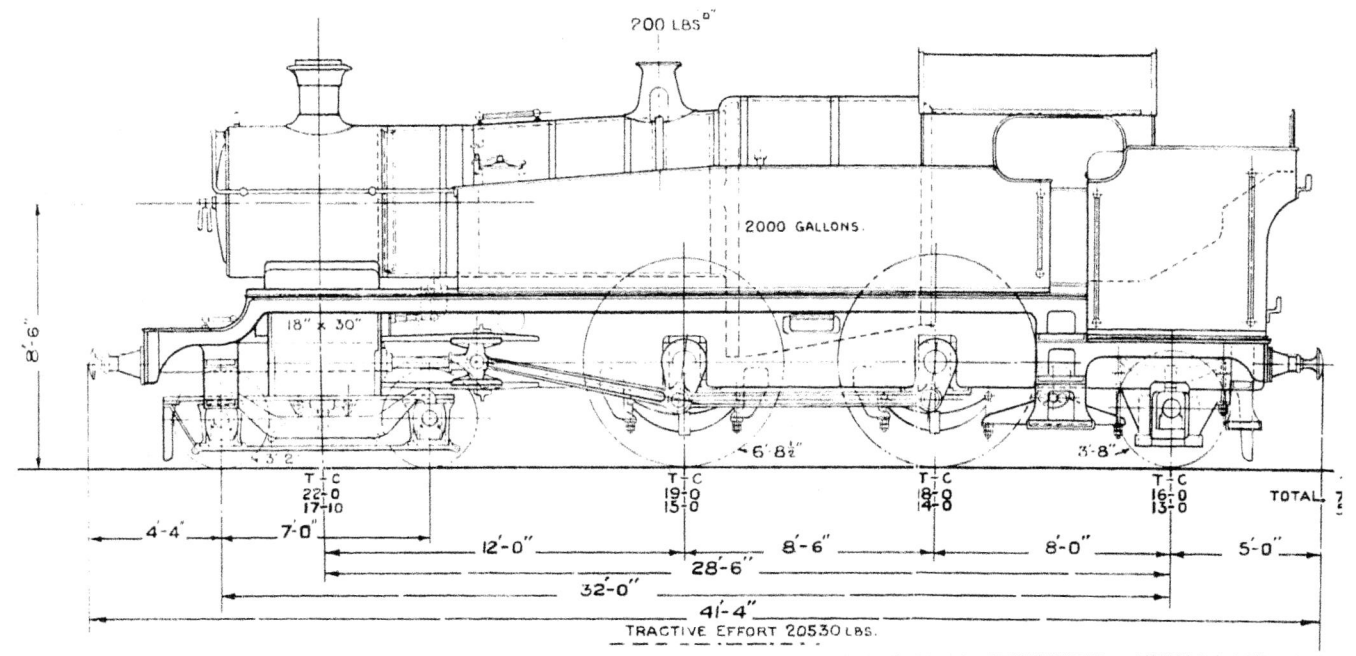

Front view

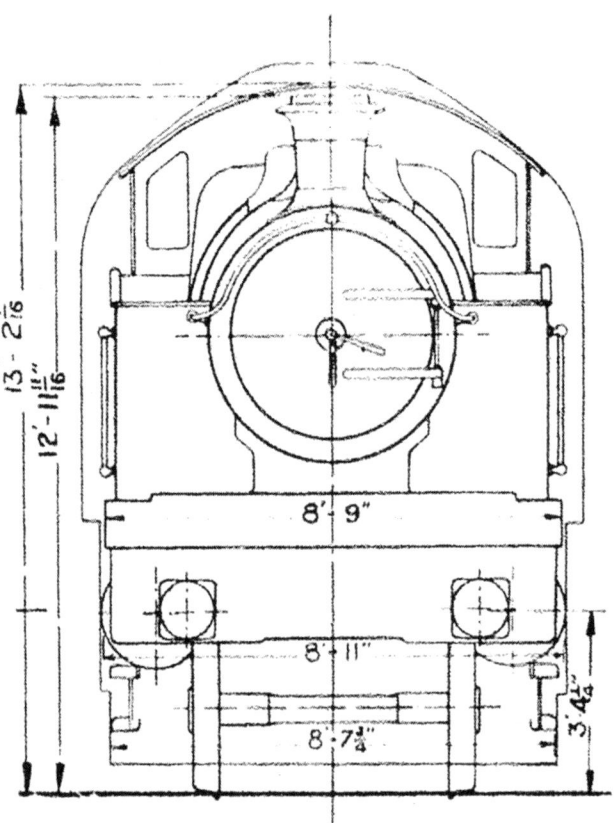

Statistics

No	Built	First Allocation Final Allocation	Withdrawn	Final Mileage
2221	9/05	Trowbridge Reading	9/33	641,239
2222	8/06	Old Oak Common Didcot	11/34	855,973
2223	9/06	Old Oak Common Old Oak Common	4/32	864,770
2224	9/06	Old Oak Common Old Oak Common	9/33	819,194
2225	9/06	Old Oak Common Carmarthen	11/34	865,958
2226	10/06	Old Oak Common Old Oak Common	6/34	875,317
2227	10/06	Old Oak Common Swindon	6/31	787,517
2228	10/06	Old Oak Common Slough	1/31	773,016
2229	10/06	Old Oak Common Reading	1/32	795,763
2230	10/06	Basingstoke Reading	9/32	752,009
2231	9/08	Reading Slough	6/31	751,851
2232	9/08	Reading Old Oak Common	10/33	841,920
2233	10/08	Reading Reading	10/32	819,127

No	Built	First Allocation Final Allocation	Withdrawn	Final Mileage
2234	10/08	Slough Old Oak Common	/32	779,383
2235	11/08	Reading Reading	1/35	801,384
2236	11/08	Reading Old Oak Common	1/32	739,757
2237	11/08	Old Oak Common Old Oak Common	6/31	804,141
2238	12/08	Old Oak Common Slough	7/31	743,578
2239	1/09	Old Oak Common Reading	11/34	872,778
2240	1/09	Old Oak Common Slough	11/31	763,113
2241	6/12	Basingstoke Swindon	5/31	582,906
2242	6/12	Old Oak Common Swindon	9/35	764,065
2243	6/12	Old Oak Common Old Oak Common	11/34	783,234
2244	6/12	Old Oak Common Reading	/33	699,909
2245	6/12	Aylesbury Slough	3/31	662,883
2246	6/12	Old Oak Common Old Oak Common	11/35	782,073
2247	7/12	Old Oak Common Reading	10/33	664,434
2248	7/12	Old Oak Common Old Oak Common	10/31	671,435
2249	8/12	Slough Reading	/32	696,360
2250	8/12	Old Oak Common Old Oak Common	2/34	726,496

4600, Churchward 4-4-2T
Weight Diagram

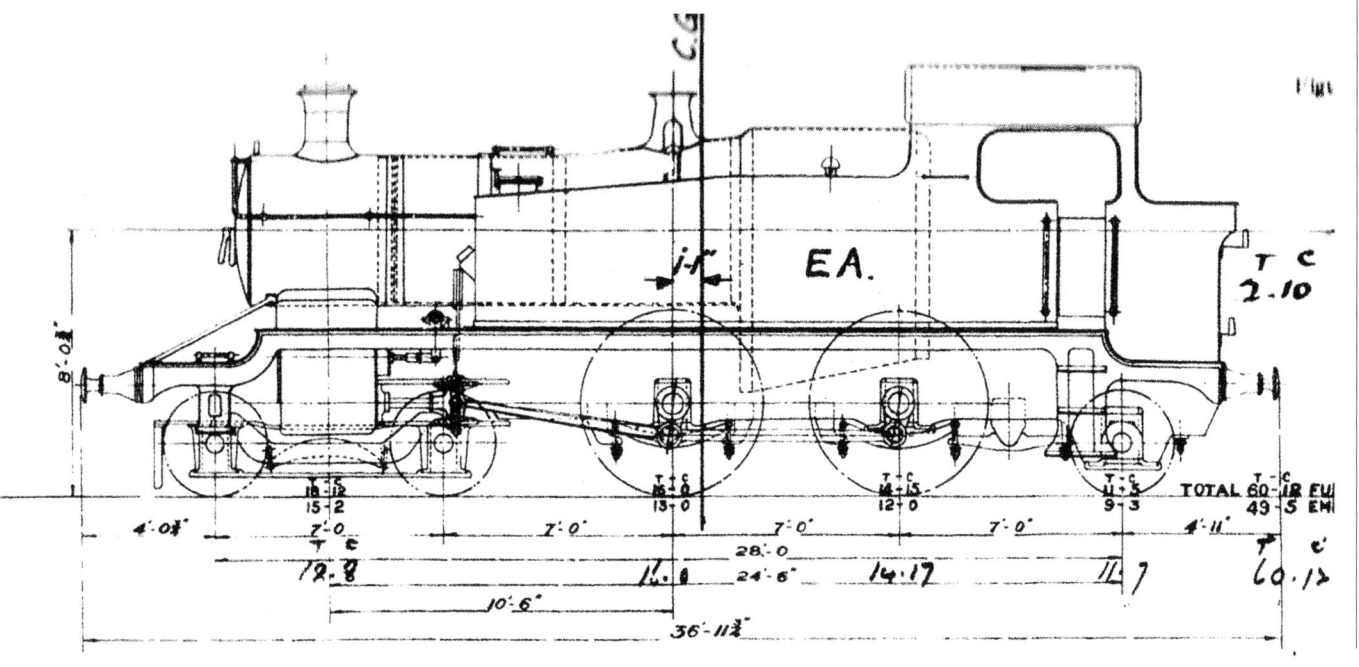

Statistics

No.	Built	First depot	Last depot	Withdrawn	Mileage
4600	11/13	Tyseley	Neyland	7/25	248,458

1-99, Steam Railmotors 0-4-0T, 1903
Statistics

All built at Swindon unless indicated'

No.	Built	Maker	First engine	Last engine	Withdrawn
1	10/1903		0801	0801	6/1917
2	10/1903		0802	0802	1/1917
3	4/1904		0803		6/1915
4	4/1904		0804		1/1915
5	4/1904		0805		5/1915
6	4/1904		0806		1/1915
7	5/1904		0809		1/1915
8	5/1904		0810		2/1915
9	5/1904		0811	0834	8/1916
10	5/1904		0812		12/1916
11	6/1904		0814		3/1917
12	6/1904		0815	0818	12/1916
13	6/1904		0816	0830	9/1922
14	7/1904		0818	0833	9/1922
15	10/1905	Kerr, Stuart	0864	0864	4/1920
16	11/1905	Kerr, Stuart	0865	0865	12/1927
17	4/1904		0807	0893	9/1922
18	4/1904		0808	0815	10/1920
19	7/1904		0820	0851	9/1922
20	8/1904		0824	0803	9/1922
21	6/1904		0817	0804	10/1920
22	7/1904		0822	0852	9/1919
23	8/1904		0823	0811	10/1919
24	7/1904		0821	0816	10/1919
25	5/1904		0813	0823	10/1919
26	6/1904		0819	0887	9/1919
27	8/1904		0825	0835	9/1919
28	9/1904		0826	0806	9/1919

No.	Built	Maker	First engine	Last engine	Withdrawn
29	1/1905		0835	0813	9/1919
30	1/1905		0833	0892	4/1935
31	12/1904		0827	0817	10/1920
32	12/1904		0828	0836	9/1919
33	12/1904		0830	0886	9/1922
34	1/1905		0832	0821	9/1922
35	12/1904		0829	0831	10/1920
36	12/1904		0831	0832	10/1920
37	2/1905		0844	0890	6/1935
38	3/1905		0848		12/1927
39	4/1905		0849	0909	6/1933
40	3/1905		0847	0884	6/1933
41	1/1905		0836	0822	8/1930
42	1/1905		0834	0807	7/1920 (Sold)
43	2/1905		0837		4/1923
44	2/1905		0839		4/1923
45	2/1905		0840	0853	2/1928
46	2/1905		0842		11/1922
47	2/1905		0843		10/1922
48	3/1905		0845		1/1916
49	2/1905		0841	0858	7/1920 (Sold)
50	3/1905		0846		4/1923
51	4/1905		0850		4/1923
52	2/1905		0838		4/1923
53	9/1905		0859	0901	6/1933
54	9/1905		0860	0910	12/1926
55	10/1905	Wolverhampton	0855	0907	5/1935
56	10/1905	Wolverhampton	0856		1/1930
57	9/1905		0861		12/1927
58	9/1905	Wolverhampton	0857	0861	6/1933
59	11/1905		0862		6/1920
60	10/1905	Wolverhampton	0853		6/1920
61	3/1906	Kerr, Stuart	0866		12/1927
62	4/1906	Kerr, Stuart	0867	0862	12/1929
63	4/1906	Kerr, Stuart	0868		12/1927

FOUR-COUPLED TANK LOCOMOTIVE CLASSES BUILT BY THE GREAT WESTERN RAILWAY

No.	Built	Maker	First engine	Last engine	Withdrawn
64	4/1906	Kerr, Stuart	0869	0863	12/1934
65	4/1906	Kerr, Stuart	0870	0842	10/1935
66	5/1906	Kerr Stuart	0871	0860	12/1934
67	6/1906	Kerr, Stuart	0872		12/1927
68	6/1906	Kerr, Stuart	0873	0859	9/1922
69	6/1906	Kerr, Stuart	0874	0885	6/1933
70	6/1906	Kerr, Stuart	0875	0896	1/1935
71	6/1906	Kerr, Stuart	0876	0846	6/1935
72	7/1906	Kerr, Stuart	0877	0844	12/1934
73	4/1906		0878	0847	6/1933
74	4/1906		0879	0900	6/1933
75	5/1906		0880	0850	12/1929
76	4/1906		0881	0877	1/1935
77	4/1906		0882	0888	12/1934
78	6/1906		0883	0849	1/1930
79	6/1906		0884	0897	5/1934
80	6/1906		0885	0848	12/1934
81	5/1907		0890	0856	12/1934
82	5/1907		0889	0841	6/1933
83	5/1907		0876	0912	6/1933
84	12/1907		0899		3/1930
85	12/1907		0900	0867	6/1914
86	12/1907		0897	0906	10/1933
87	12/1907		0898	0903	5/1928
88	1/1908		0902	0899	10/1935
89	12/1907		0903		12/1927
90	12/1907		0901		12/1927
91	1/1908		0905	0882	1/1935
92	1/1908		0904	0839	10/1935
93	2/1908	Wolverhampton	0906	0873	12/1934
94	2/1908	Wolverhampton	0907	0898	10/1929
95	2/1908		0908		3/1930
96	2/1908		0910	0869	12/1934
97	2/1908		0912	0872	4/1935
98	2/1908		0909	0908	5/1935
99	2/1908		0911	0895	5/1928

0931-0942 were built at Wolverhampton as 'spares' and some of the 'blanks' above were either as built (especially the early ones) or these 'spares'.

Experimental locomotives
1, Dean 4-4-0T, 1880, reb.2-4-0T, 1882

13, Dean 2-4-2T, 1886, reb.4-4-0T, 1897
Weight diagram

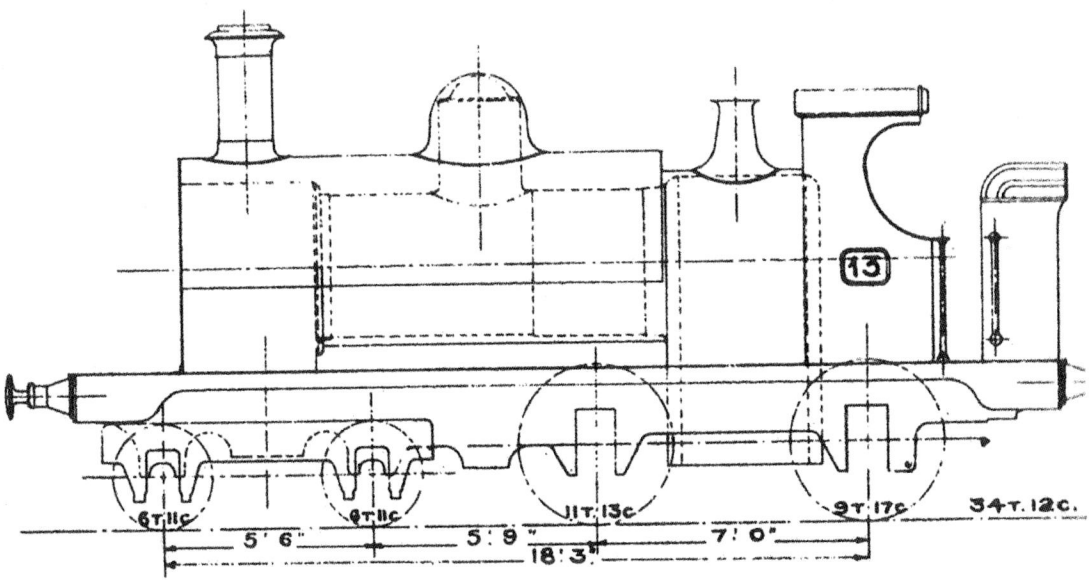

34, 35, Dean 0-4-2ST, 1890, reb 0-4-4T, 1895
Weight diagrams

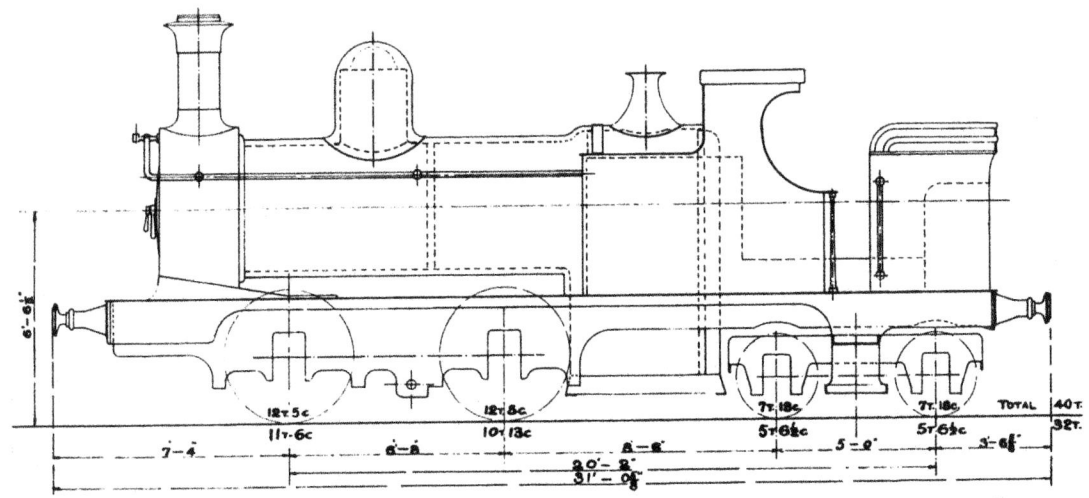

1490, 4-4-0PT
Weight diagram

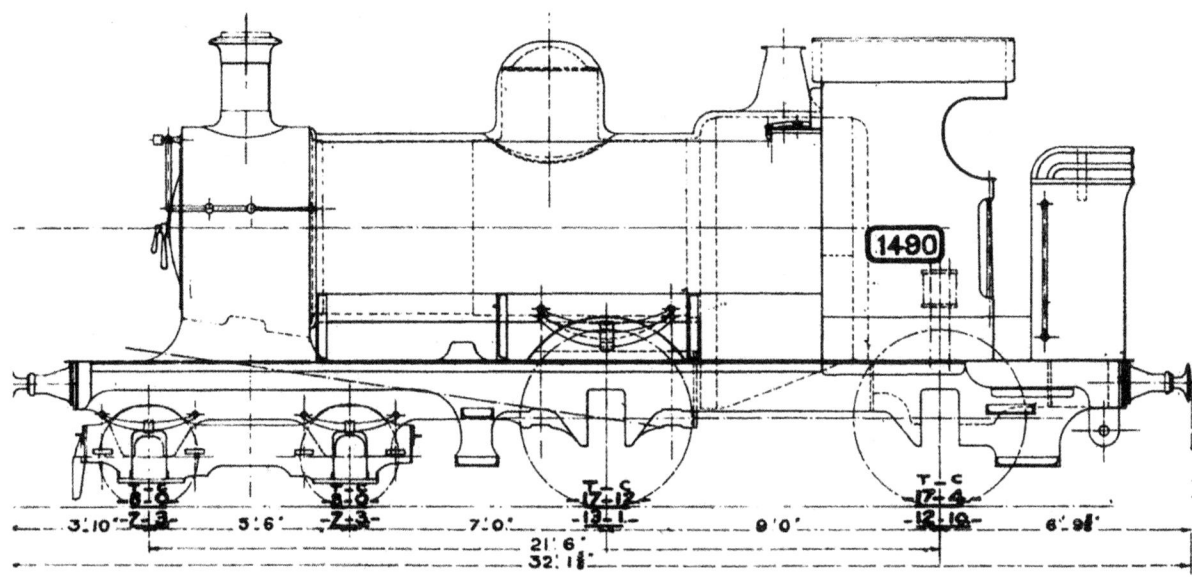

101, oil burning 0-4-0T, 1902
Weight diagram

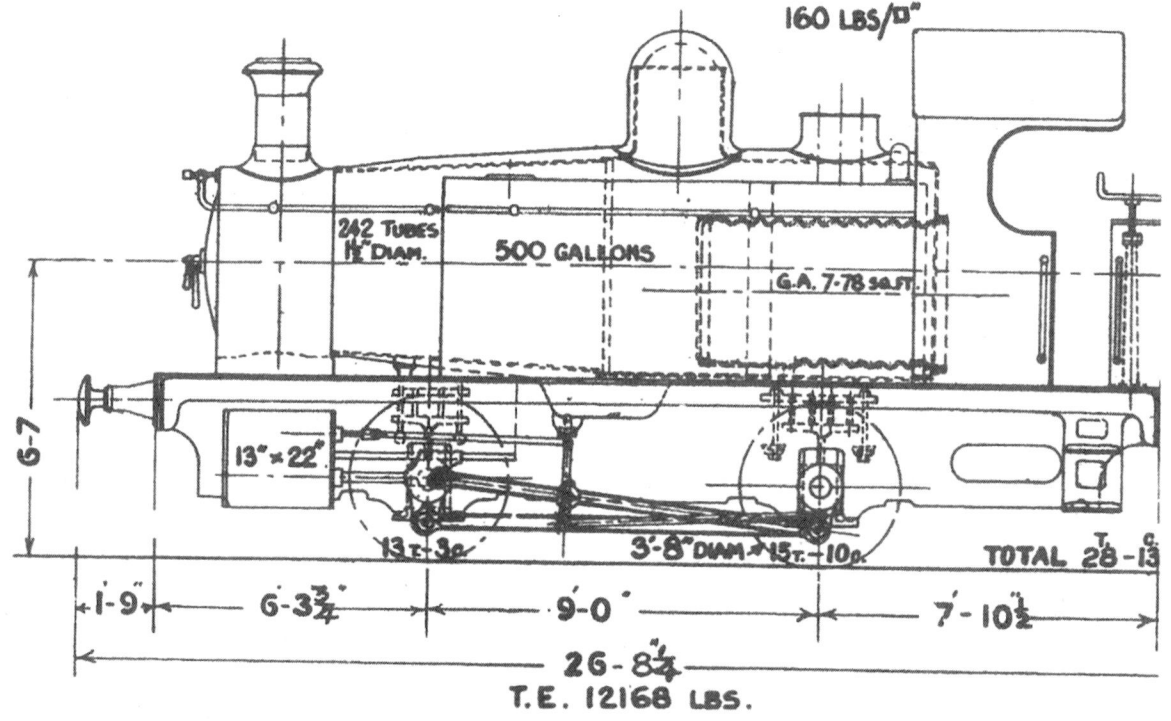

Statistics

No.	Built	Rebuilt	Withdrawal	Remarks
1	5/1880	5/1882	7/1924	
13	10/1886	12/1897	5/1926	
34	1/1890	10/1895	9/1908	
35	1/1890	11/1895	3/1906	
101	6/1902	5/1905	9/1911	
1490	10/98		11/1907	Sold to Ebbw Vale Steel Co

GWR classes post-1923
1101-1106, 0-4-0T, 1926
Weight diagram

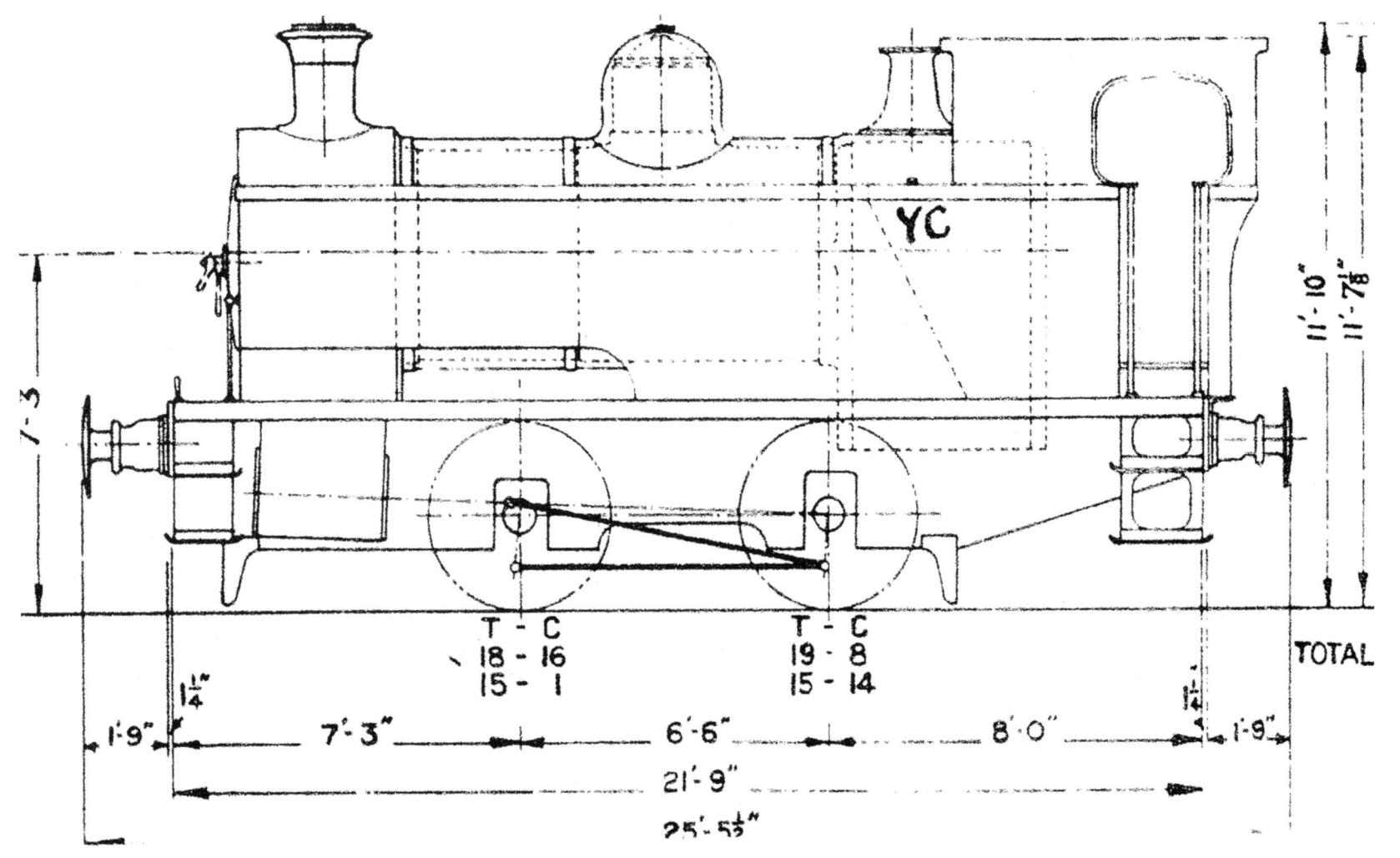

Statistics

No.	Built	Withdrawal
1101	6/1926	11/1959
1102	7/1926	2/1960
1103	7/1926	2/1960
1104	8/1926	2/1960
1105	8/1926	2/1960
1106	8/1926	2/1960

12-13, Sentinel Waggon Works 0-4-0T, 1926
Statistics

No.	Built	Withdrawal	Remarks
12	1926	12/1926	Returned to maker & sold
13	1926	5/1946	Sold

4800-4874 (renumbered 1400-1474), 0-4-2T, 1932
Weight diagram

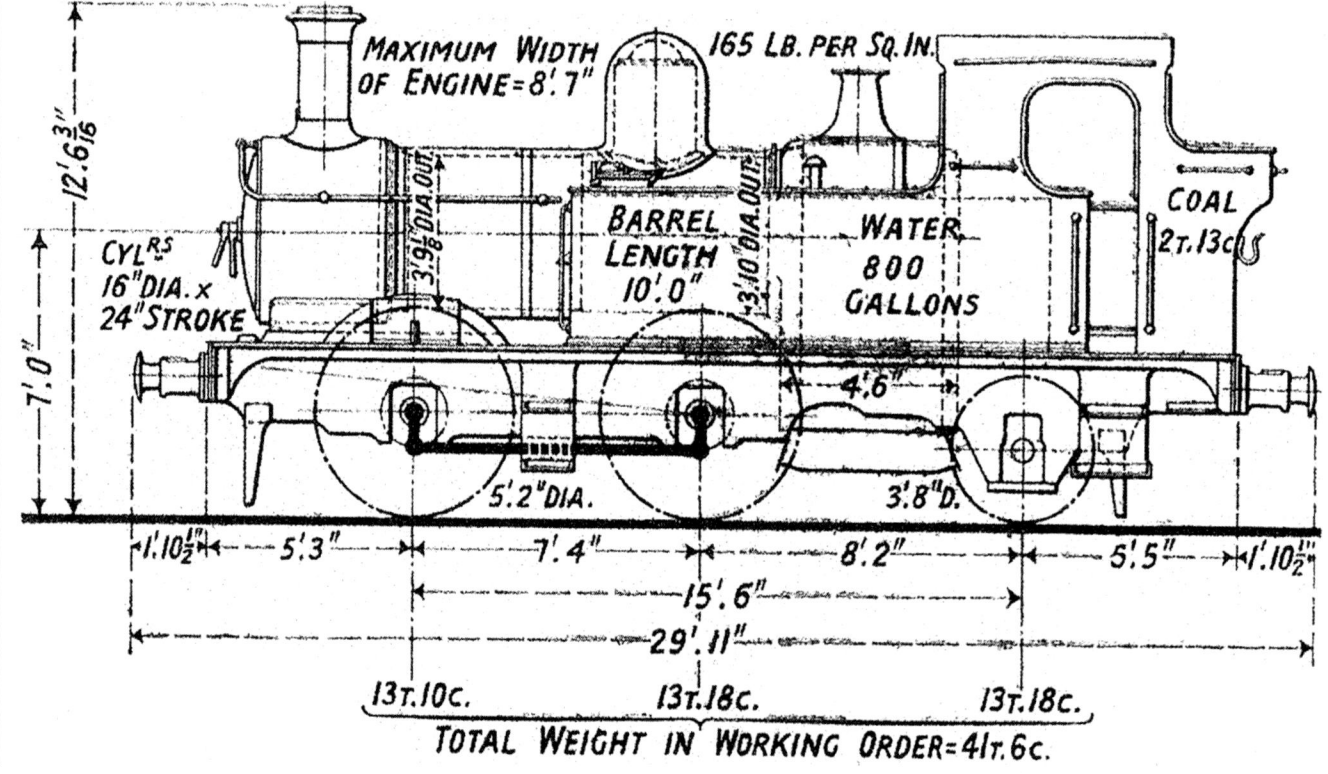

Front & rear view

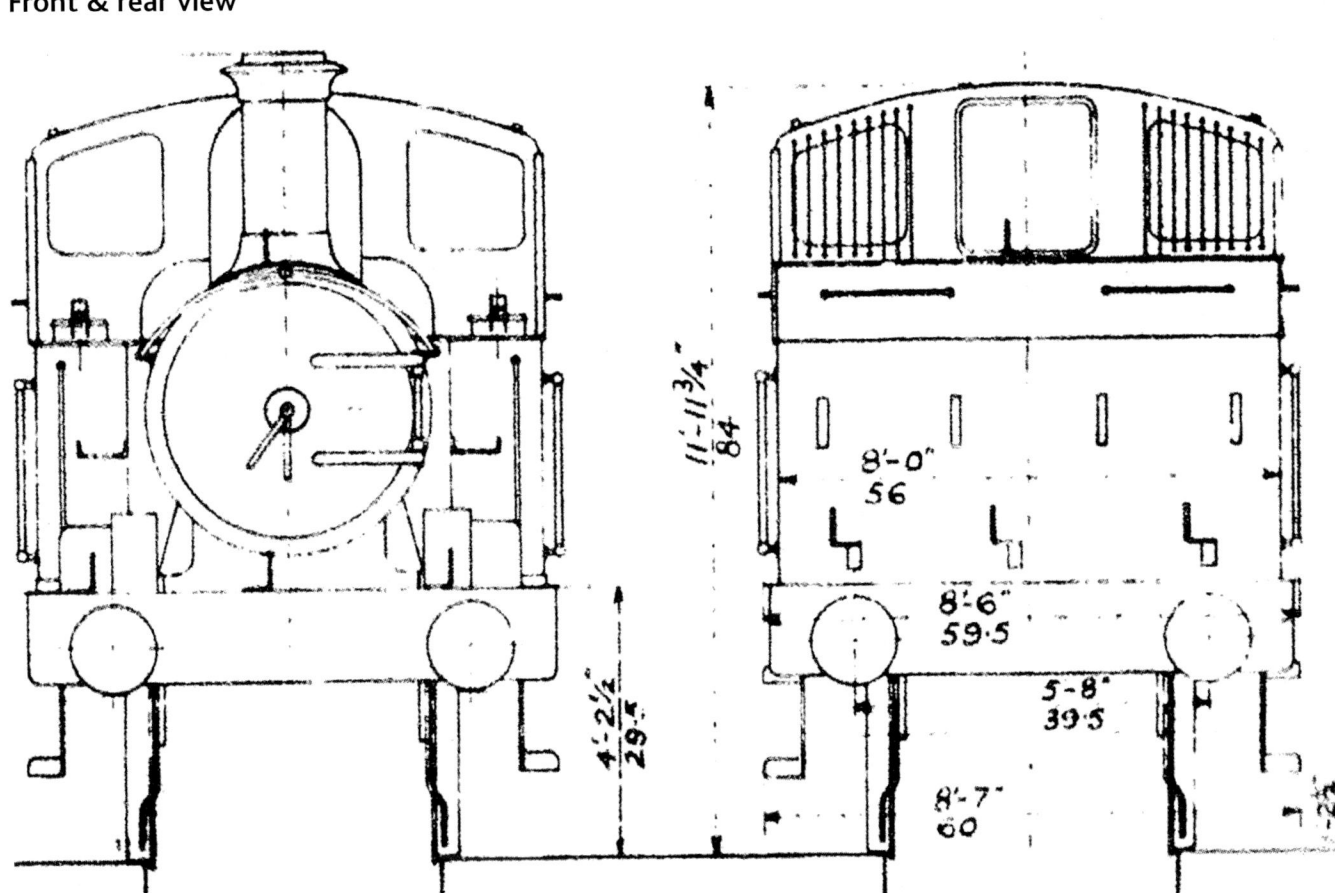

Statistics

4800-4874 renumbered 1400-1474 between 10/1946 and 12/1946.

No.	Built	1950 depot	1960 depot	Last depot	Withdrawal
1400	8/1932	Swindon		Swindon	6/1957
1401	8/1932	Banbury		Gloucester	11/1958
1402	8/1932	Gloucester		Bristol Bath Road	10/1956
1403	9/1932	Swindon		Weymouth	11/1957
1404	9/1932	Gloucester		Gloucester	2/1956
1405	9/1932	Exeter		Exeter	9/1958
1406	9/1932	Gloucester		Southall	2/1958
1407	9/1932	Reading	Reading	Southall	6/1960
1408	9/1932	Worcester		Laira	2/1958

No.	Built	1950 depot	1960 depot	Last depot	Withdrawal
1409	10/1932	Gloucester	Gloucester		10/1963
1410	3/1933	Wolverhampton	Swindon		
1411	3/1933	Banbury		Slough	10/1956
1412	3/1933	Oswestry	Bristol Bath Road	Bristol Bath Road	6/1960
1413	4/1933	Gloucester		Gloucester	3/1956
1414	4/1933	Stourbridge		Stourbridge	3/1957
1415	4/1933	Bristol Bath Road		Southall	2/1957
1416	4/1933	Croes Newydd		Croes Newydd	10/1956
1417	4/1933	Birkenhead		Swindon	1/1959
1418	4/1933	Worcester		Weymouth	10/1956
1419	4/1933	St Blazey	St Blazey		
1420	11/1933	Cardiff Cathays	Laira		1964 P
1421	11/1933	Llantrisant	Reading		12/1963
1422	11/1933	Pontypool Road		Swindon	6/1957
1423	11/1933	Goodwick		Oswestry	1/1959
1424	11/1933	Gloucester	Gloucester		11/1963
1425	11/1933	Cardiff Cathays		Oxford	2/1956
1426	11/1933	Neasden	Gloucester		4/1962
1427	11/1933	Newton Abbot	Gloucester	Gloucester	6/1960
1428	11/1933	Oswestry		Gloucester	6/1959
1429	11/1933	Exeter		Exeter	3/1959
1430	7/1934	Bristol Bath Road		Gloucester	9/1958
1431	7/1934	Goodwick	Gloucester		
1432	7/1934	Oswestry	Oswestry		7/1963
1433	7/1934	Swindon	Gloucester		1/1961
1434	7/1934	Chester	Laira		7/1962
1435	8/1934	Exeter	Oxford		
1436	8/1934	Swindon		Southall	10/1958
1437	8/1934	Slough		Oxford	2/1959
1438	8/1934	Stourbridge	Oswestry		2/1963
1439	8/1934	Newton Abbot		Newton Abbot	8/1957
1440	3/1935	Exeter	Exeter		12/1963
1441	4/1935	Gloucester	Gloucester	Gloucester	6/1960
1442	4/1935	Slough	Oxford	Exmouth Junction	5/1965 P
1443	4/1935	Southall		Southall	6/1957

No.	Built	1950 depot	1960 depot	Last depot	Withdrawal
1444	4/1935	Reading	Oxford		1964
1445	4/1935	Hereford	Hereford	Gloucester	10/1964
1446	4/1935	Swindon		Southall	9/1958
1447	4/1935	Reading	Slough		4/1964
1448	4/1935	Oxford	Slough	Slough	6/1960
1449	4/1935	Exeter	Machynlleth	Machynlleth	6/1960
1450	7/1935	Oxford	Oxford	Exmouth Junction	5/1965 P
1451	7/1935	Exeter	Exeter		7/1964
1452	7/1935	Goodwick	Exeter	Exeter	6/1960
1453	7/1935	Weymouth	Slough		1964
1454	7/1935	Weymouth	Gloucester		12/1960
1455	7/1935	Hereford	Hereford		7/1964
1456	7/1935	Gloucester		Hereford	2/1959
1457	8/1935	Croes Newydd		Machynlleth	2/1959
1458	8/1935	Banbury	Oswestry		1964
1459	8/1935	Oswestry		Weymouth	9/1958
1460	2/1936	Hereford		Hereford	2/1956
1461	2/1936	Cardiff Cathays		Gloucester	5/1958
1462	2/1936	Southall	Exeter		9/1962
1463	2/1936	Bristol Bath Road	Bristol Bath Road		
1464	2/1936	Gloucester	Swindon	Swindon	6/1960
1465	2/1936	Machynlleth		Croes Newydd	9/1958
1466	2/1936	Newton Abbot	Newton Abbot		12/1963 P
1467	2/1936	Weymouth		Gloucester	4/1959
1468	2/1936	Exeter	Exeter		3/1962
1469	2/1936	Exeter		Exeter	9/1958
1470	4/1936	Newton Abbot	Newton Abbot		10/1962
1471	4/1936	Llantrisant	Exeter		10/1963
1472	4/1936	Carmarthen	Gloucester	Gloucester	1964
1473	4/1936	Croes Newydd	Neasden		8/1962
1474	4/1936	Machynlleth	Southall		10/1964

P = Preserved

5800-5819, 0-4-2T, 1933
Weight diagram as for 4800 class
Statistics

No.	Built	1950 depot	Last depot	Withdrawal
5800	1/1933	Swindon	Swindon	7/1958
5801	1/1933	Brecon	Machynlleth	9/1958
5802	1/1933	Swindon	Swindon	12/1958
5803	1/1933	Oswestry	Machynlleth	7/1957
5804	1/1933	Swindon	Swindon	6/1959
5805	1/1933	Swindon	Swindon	3/1958
5806	1/1933	Oswestry	Oswestry	6/1957
5807	1/1933	Hereford	Hereford	6/1957
5808	1/1933	Hereford	Oxford	2/1957
5809	2/1933	Bristol Bath Road	Machynlleth	8/1959
5810	8/1933	Croes Newydd	Croes Newydd	1/1959
5811	8/1933	Croes Newydd	Oxford	5/1957
5812	8/1933	Oswestry	Oswestry	6/1957
5813	8/1933	Bristol Bath Road	Leamington Spa	11/1957
5814	8/1933	Hereford	Hereford	6/1957
5815	8/1933	Worcester	Swindon	4/1963
5816	8/1933	Worcester	Oxford	7/1957
5817	8/1933	Hereford	Hereford	6/1957
5818	8/1933	Pontypool Road	Oxford	9/1959
5819	8/1933	Carmarthen	Carmarthen	6/1957

BIBLIOGRAPHY

Casserley, H.C., *Locomotives at the Grouping, No.4 Great Western Railway,* Ian Allan, 1966
Freezer, C.J., *Locomotives in Outline, GWR,* Peco Publications, 1977
Maidment, D.J., *Great Western County Classes,* Pen & Sword, 2018
Maidment, D.J., *Great Western Pannier Tank Classes,* Pen & Sword, 2019
Maidment, D.J., *Great Western Small-wheeled Double-framed 4-4-0 Tender Locomotives,* Pen & Sword, 2017
Pritchard, Robert & Hall, Peter, *Preserved Locomotives of British Railways,* Platform 5, 2016
RCTS, *The Locomotives of the Great Western Railway, Parts 1-7,* RCTS, 1951
RCTS, *The Locomotives of the Great Western Railway, Parts 8-12,* RCTS, 1953
Russell, J.H., *A Pictorial Record of Great Western Engines, Volume 1,* Oxford Publishing Co., 1975
X-press, *The Xpress Locomotive Register, Volume 4. Western Region, 1950-1960,* Xpress Publishing
Wilson, Andrew, 'Collett 4800 and 5800 class 0-4-2Ts of the Great Western Railway', article in *Steam Days,* 2006.

INDEX

Bibliography, 195
Engineers, 9-11
    Armstrong, George, 10
    Armstrong, Joseph, 9
    Churchward, George Jackson, 11
    Collett, Charles Benjamin, 11-12
    Dean, William, 10-11
    Gooch, Sir Daniel, 9
Locomotive classes (Broad Gauge)
Convertible, 3501 class 2-4-0T, 17-18
    Construction, 18
    Dimensions, 18
    Operations, 18
    Rebuilding, 18
    Statistics, 163-164
Convertible, 3541 class 0-4-2ST/0-4-4T, 18-21
    Construction, 18
    Dimensions, 19
    Operations, 18-19
    Rebuilding, 18-19
    Statistics, 164
    Withdrawal, 19
Corsair 4-4-0ST, 14-15
    Builders, 14
    Dimensions, 14
    Drawing, 14
    Mileage, 15
    Names, 14
    Operations, 15
    Statistics, 162
    Withdrawal, 15
Hawthorn 2-4-0ST, 16-17
    Builders, 16
    Dimensions, 16
    Names, 16
    Operations, 16-17
    Statistics, 163
    Withdrawal, 17
Leo 2-4-0T, 13-14
    Builders, 14
    Dimensions, 14
    Drawing, 13
    Names, 14
    Operations, 14
    Statistics, 161
    Withdrawal, 14
Metropolitan Tanks, 2-4-0T, 15-16
    Builders, 15
    Dimensions, 15
    Names, 15
    Operations, 15
    Statistics, 162-163
    Withdrawal, 15-16
Locomotive classes (Standard Gauge)
45, 0-4-0ST, 25-26
    Construction, 25
    Dimensions, 26
    Mileage, 26
    Operation, 26
    Statistics, 165
    Withdrawal, 26
91/92, Beyer Peacock 0-4-2ST/0-4-0ST, 22-24
    Builders, 22
    Dimensions, 22
    Mileage, 24
    Operations, 24
    Weight diagram, 165
    Withdrawal, 24
117, 4-4-0T, 46
    Construction, 46
    Dimensions, 46
    Statistics, 172
    Withdrawal, 46
320/321, 46-47
    Condensing gear, 46
    Construction, 47
    Dimensions, 47
    Drawing, 46
    Mileage, 47
    Operation, 47
    Statistics, 172
    Withdrawal, 47
342/343, Beyer Peacock 0-4-0ST, 24-25
    Builders, 25
    Dimensions, 25
    Mileage, 25
    Operations, 25
    Statistics, 165
    Withdrawal, 25
344 class, 2-4-0T, 25-27
    Builders, 25-26
    Dimensions, 25-26
    Mileage, 25
    Numbers, 26
    Operations, 27
    Statistics, 166
    Withdrawal, 25, 27
455 class 'Metro Tank' 2-4-0T, 47-66
    ATC, 47
    Auto-working, 59
    Condensing gear, 47, 50
    Construction, 47, 50, 54
    Dimensions, 47
    Modifications, 50, 54
    Operations, 58-59
    Statistics, 173-177
    Weight diagrams, 172-173
    Withdrawal, 47, 50, 54, 59

Index

517 class, 0-4-2T, 27-43
    Construction, 27-28
    Dimensions, 27-28
    Modifications, 278
    Livery, 30
    Operations, 35
    Statistics, 166-171
    Weight diagrams, 166
    Withdrawal, 27, 35
1101 class, 0-4-0PT, 94-96
    Construction, 94, 96
    Dimensions, 96
    Operations, 96
    Statistics, 190
    Weight diagram, 189
    Withdrawal, 96
2221 class 'County Tank' 4-4-2T, 77-81
    Allocation, 79-80
    Construction, 77
    Dimensions, 77-78
    Drawing for 4-4-4T, 78
    Mileage, 79
    Modifications, 78
    Livery, 78
    Operations, 79-81
    Statistics, 182-183
    Superheating, 78
    Weight diagrams, 181-182
    Withdrawal, 78-79
3511 class, 2-4-0T, 67
    Construction, 67
    Dimensions, 67
    Operations, 67
    Statistics, 178
    Rebuilding, 67
3521 class, 0-4-2T/0-4-4T, 68-69
    Construction, 68
    Derailments, 69
    Dimensions, 68
    Operations, 68
    Statistics, 178
    Rebuilding, 68
3571 class, 0-4-2T, 44-46
    Construction, 44
    Dimensions, 44
    Mileage, 46
    Modifications, 44
    Operations, 46
    Statistics, 171-172
    Weight diagram, 171
    Withdrawal, 44, 46
3600 class, 2-4-2T, 69-76
    ATC fitting, 70
    Construction, 69
    Dimensions, 69-70
    Mileage, 70
    Modifications, 70
    Operations, 70
    Statistics, 180
    Weight diagrams, 179
    Withdrawal, 70
4600 class 4-4-2T, 82-83
    Construction, 82
    Dimensions, 82-83
    Statistics, 184
    Weight diagram, 183
48XX (14XX) 0-4-2T, 97-127
    Allocation, 104-105, 125, 127
    ATC, 104
    Auto-fitted, 104
    Construction, 97
    Costs, 97
    Dimensions, 98
    Filming, 105
    Mileage, 104, 127
    Livery, 98
    Logs, 106, 107
    Operations, 104-107
    Performance, 104, 106, 107
    Personal experience, 123, 125
    Preservation, 136-143
    Statistics, 191-193
    Weight diagrams, 190-191
    Withdrawal, 125, 127
58XX 0-4-2T, 98, 128-135
    Allocation, 128
    ATC, 128
    Construction, 128
    Dimensions, 98
    Operations, 128
    Performance, 128
    Statistics, 194
    Withdrawal, 128
Experimental Loco No.1, 4-4-0T/2-4-0T, 87-88
    Construction, 87
    Dimensions, 87-88
    Drawing as 4-4-0T, 87
    Mileage, 88
    Operations, 88
    Rebuilding, 87-88
    Statistics, 189
    Withdrawal, 88
Experimental Loco No.13, 2-4-2T/4-4-0T, 88-89
    Construction, 88
    Dimensions, 88-89
    Mileage, 89
    Operations, 89
    Rebuilding, 89
    Statistics, 189
    Weight diagram, 187
    Withdrawal, 89
Experimental Locos Nos.34, 35, 0-4-2T/0-4-4T, 90-91
    Construction, 90
    Dimensions, 90
    Operations, 91
    Rebuilding, 90
    Statistics, 189
    Weight diagram, 187
    Withdrawal, 91
Experimental Locos No.101, 0-4-0T, oil-burning, 93
    Construction, 93
    Dimensions, 93
    Operations, 93
    Rebuilding, 93
    Statistics, 189
    Weight diagram, 188
    Withdrawal, 93
Experimental Locos No.1490, 4-4-0PT, 91-93
    Construction, 91
    Dimensions, 91
    Operations, 91
    Statistics, 189
    Weight diagram, 188
    Withdrawal, 91
Sentinel Nos. 12, 13, 96-97

Builders, 96
Construction, 96
Dimensions, 97
Operations, 97
Preservation, 143
Sale/Withdrawal, 97
Statistics, 190
Steam Railmotors, 0-4-0T, 83-86
Builders, 83
Construction, 83
Dimensions, 83
Livery, 83
Operations, 83-84
Preservation, 144
Statistics, 184-187
Withdrawal, 83

Photos, Location (black & white)
Abingdon, 123
Ashburton, 112, 138
Bala branch, 134
Barmouth, 131
Bearley, 109
Birkenhead Woodside, 75
Birmingham Snow Hill, 39, 71
Blaenau Ffestiniog, 133
Blenheim & Woodstock, 42, 108
Brimscombe, 84, 118
Bristol Temple Meads, 40, 62
Buckfastleigh, 137, 141
Builth Road, 108
Builth Wells, 130
Burton Latimer, 143
Caerphilly Works, 127
Calne, 106
Capenhurst, 75
Cardiff Canton, 103
Cardigan, 37
Chalford, 84, 126
Chard, 63
Cheltenham St James, 60
Chester, 44, 45, 88
Cholsey, 143
Churston, 122
Cinderford, 116
Cramlington Colliery, 92
Croes Newydd, 24
Danygraig, 95

Dean Forest Workshops, 140
Didcot GW Railway Centre, 140-142, 144
Dinas Mawddwy, 39
Dulverton, 110
Ealing Broadway, 125
Elson, 118
Exeter St David's, 41, 100
Fowey, 117
Gloucester, 70, 99
Hannington, 132
Hatton, 42
Hayes, 81
Heathfield, 111
Hemyock, 114, 115
Hereford, 136
Highworth, 131
Hooton, 46, 76
Kensal Green, 65
Lamphey, 86
Leamington Spa, 76
Llandyssll, 135
Longmoor, 90
Looe, 89
Machynlleth, 103
Malmesbury, 132
Marlow, 109, 124
Monkton Combe, 104
Monmouth Troy, 115
Moretonhampstead, 110
New Tredegar, 92
Newton Abbot, 63, 102
Old Oak Common, 52, 60, 73
Oswestry, 40, 101, 117
Oxford, 35, 64, 66
Paddington, 55, 61, 62, 64, 80, 123
Penrhyn, 59
Penygraig, 120
Plymouth North Road, 69
Pontrilas, 129
Porthygwaen, 113
Radyr, 23
Radley, 119
St Fagan's, 74
Saltney Junction, 43, 121
Severn Tunnel Junction, 24, 67
Slough, 57, 62, 92

Southall, 41, 57, 86
Staines, 41
Stourbridge Junction, 74
Sudeley, 86
Swansea East Dock, 96
Swindon, 31, 53, 56, 74, 89, 94, 99, 126, 129, 130, 135, 136
Symond's Yat, 116
Talyllyn, 133
Tetbury, 134
Titfield Thunderbolt filming, 104, 105
Tiverton, 139
Toddington, 140
Totnes, 101, 112
Trumpers Crossing (Osterley Park), 37, 38
Truro, 65
Tyseley, 82
Uffculme, 121, 122
Upton Scudamore, 113
Wallingford, 119
Weymouth, 124
Wolverhampton Stafford Road, 23, 27, 29-31, 33, 72, 102
Worcester, 63, 138
Yeovil, 43

Photos, Location (colour)
Ashburton, 151, 152
Bampton, 156
Bourne End, 147
Chalford, 150
Chinnor, 160
Churston, 154
Cowley Bridge Junction, 154
Dartington, 153
Dulverton, 159
Exeter St David's, 160
Fairby, 156
Hemyock, 154, 155
Hereford, 147
Lostwithiel, 160
Marlow, 146, 147
Monmouth Troy, 149
Moretonhampstead, 152
Princes Risborough, 148
Radley, 148

Sharpness, 151
Southall, 145
Stroud, 150
Swindon, 146
Symond's Yat, 149
Tiverton, 153, 157, 158
Tiverton Junction, 159
Thorverton, 158
Yatton, 148

Photos – Locomotives, Broad Gauge (black & white)
3507, Broad Gauge Convertible, 17
3509, Broad Gauge Convertible, 18
3541, Broad Gauge Convertible, 19
3547, Broad Gauge Convertible, 21
3548, Broad Gauge Convertible, 20
3554, Broad Gauge Convertible, 21
3557, Broad Gauge Convertible, 19
3560, Broad Gauge Convertible, 20
Hedley, Broad Gauge, 17
Locust, Broad Gauge, 15
Melling, Broad Gauge, 16
Ostrich, Broad Gauge, 16

Photos – Locomotives, Standard Gauge (black & white)
3, 455 cl, 59
5, 455 cl, 48, 63
6, 455 cl, 61
11, 36XX class, 71
45, 0-4-0ST, 26
92, 0-4-0ST, 23-25
202, 517 cl, 37
204, 517 cl, 29, 40
205, 517 cl, 33, 41
342, 0-4-0ST, 24, 25
346, 2-4-0ST, 27
455, 455 cl 2-4-0T, 48
457, 455 cl, 47, 49, 66
515, 517 cl, 0-4-2ST, 28
517, 517 cl, 32
518, 517 cl, 38
519, 517 cl, 32
530, 517 cl, 31, 42
538, 517 cl, 36
547, 517 cl, 41
551, 517 cl, 31
555, 517 cl, 42

556, 517 cl, 0-4-2ST, 28
565, 517 cl, 32
567, 517 cl, 0-4-2ST, 28
569, 517 cl, 0-4-2ST, 36
574, 517 cl, 40
626, 455 cl, 50
627, 455 cl, 60
632, 455 cl, 49, 62
830, 517 cl, 42
832, 517 cl, 39
833, 517 (as railmotor disguise), 38
833, 517 cl, 41
834, 517 cl, 30
837, 517 cl, 39
967, 455 cl, 51
974, 455 cl, 50
975, 455 cl, 63
980, 455 cl, 63
985, 455 cl, 58
1101 class, 96
1104, 94
1105, 95
1154, 517 cl, 33
1155, 517 cl, 39
1157, 517 cl, 43
1164, 517 cl, 29
1165, 517 cl, 37
1401, 104, 105, 116, 126
1404, 116, 127
1406, 455 cl, 52
1407, 455 cl, 62
1410, 106
1411, 455 cl, 60
1412, 113
1419, 117
1420, 123, 136-138
1421, 517 cl, 30, 31
1423, 126
1424, 102, 118
1426, 103
1428, 517 cl, 43
1428, 117
1429, 114, 115
1432, 101, 118
1438, 517 cl, 34
1439, 111
1442, 139

1447, 119, 123, 124
1450, 455 cl, 52
1450, 140, 141
1453, 455 cl, 53
1453, 124
1455, 115
1458, 120, 125
1461, 455 cl, 60
1462, 121
1463, 102
1465, 517 cl, 43
1465, 103, 121
1466, 517 cl, 34, 40
1466, 143
1468, 122
1469, 127
1470, 112, 122
1471, 120
1472, 517 cl, 34
1472, 126
1473, 517 cl, 35
1494, 455 cl, 53
1498, 455 cl, 64
1499, 455 cl, 65
2038, 0-6-0PT, 65
2227, 77
2235, 80
2240, 81
3440 (3717), 140
3512, 67
3516, 67
3527, 69
3537, 68
3561, 455 cl, 54
3565, 455 cl, 61
3568, 455 cl, 54
3569, 455 cl, 64
3570, 455 cl, 62
3571, 46
3575, 45
3576, 44
3578, 46
3579, 45
3583, 455 cl, 55
3585, 66
3587, 455 cl, 61
3593, 455 cl, 56

3595, 455 cl, 56
3596, 455 cl, 57, 65
3600 (ex-11), 73
3601, 64, 73
3604, 71
3612, 74
3615, 74
3616, 72, 76
3620, 72
3622, 73
3625, 75
3627, 75
3628, 74
3629, 75, 76
4555, 138
4600, 82
4800, 98
4803, 99
4805, 98, 100
4808, 99
4816, 109
4827, 110
4841, 111
4843, 108
4847, 109
4848, 100
4849, 110
4851, 100
4866, 142
4870, 101
4874, 108
5800, 130, 132
5801, 130, 131, 133
5802, 132
5804, 135
5805, 131
5806, 128
5810, 133, 134
5811, 129
5815, 134,, 135
5818, 129
5819, 135
5976, 122
Experimental Loco No.1, 87, 88
Experimental Loco, No.13, 88, 89
Experimental Locos, Nos. 34/35, 90
Experimental Loco, No.101, 93
Experimental Loco, No.1490, 91-93
Sentinel, No.12, 143
Sentinel, No.13, 97
Steam Railmotor 1, 84
Steam Railmotor 48, 86
Steam Railmotor 59, 84
Steam Railmotor 68, 85
Steam Railmotor 70, 86
Steam Railmotor 71, 85
Steam Railmotor 93, 144
Photos – Locomotives, Standard Gauge (colour)
1409, 146
1419, 160
1421, 155
1440, 148
1442, 160
1444, 148
1445, 146, 147
1446, 145
1447, 147
1450, 156
1451, 155, 157, 158
1453, 146 150
1455, 149
1456, 149
1462, 157, 158
1463, 148
1466, 152-154, 157
1470, 151, 153, 154
1471, 157-159
1472, 150, 151
1473, 160
Swindon Works, 22
Wolverhampton Works, 22